Perspectives on Systems Engineering

from Florida's Space Coast

Edited by Scott Tilley, Ph.D.

Perspectives on Systems Engineering from Florida's Space Coast

Cover design © Scott Tilley
Cover photo © Steve Ratts
Back photo © SpaceX, Wikimedia

Published by CTS Press

CTS
Press

An imprint of Precious Publishing, LLC

Precious Publishing
www.PreciousPublishing.biz

ISBN-13: 978-1-951750-04-6

ISBN-13: 978-1-951750-05-3 (ebook)

TABLE OF CONTENTS

Dedication

To all the systems engineers on Florida's Space Coast.

Image from https://www.visitspacecoast.com

Preface

The International Council on Systems Engineering (INCOSE) defines systems engineering as "a transdisciplinary and integrative approach to enable the successful realization, use, and retirement of engineered systems, using systems principles and concepts, and scientific, technological, and management methods." This is a complex definition, but at its heart are systems engineers – the people who make it all work.

I live on the Space Coast, a distinctive area of east central Florida that is a narrow 72-mile stretch of Atlantic shoreline from Canaveral National Seashore to Sebastian Inlet. As the name suggests, the Space Coast is where the nation's space program began – and continues to thrive. I've seen Space Shuttle launches and landings, huge rockets taking off, and experienced the awesome scene of SpaceX's Falcon 9 boosters returning to earth. It's the only place I know where local hotels put three items on their beachside information boards: the air temperature, the ocean temperature, and the next launch time.

The Space Coast is also exceptional because of the large number of aerospace companies and defense contractors in the area. These organizations have some of the highest concentrations of talented systems engineers in the nation, and this book captures a few of their unique insights into the profession. Enjoy!

Scott Tilley

Melbourne, FL

September 9, 2020

STS-87 (1997): Like a rising sun lighting up the afternoon sky, Space Shuttle Columbia soars from KSC Launch Pad 39B. (NASA)

Acknowledgments

Organizations like INCOSE and the IEEE Systems Council have done a wonderful job of advancing systems engineering as a profession. Without all of the volunteers who make their projects possible, systems engineering would not have matured as fast as it has in the past few decades.

Thanks for everyone who submitted their work for consideration. Without them, this book would not exist. Thanks as well to the reviewers who diligently read and provided feedback on all the submissions. As editor, any remaining errors in the text are purely my responsibility.

A special note of thanks to Steve Ratts, who took the photograph on the cover of the book. It is a closeup of the USSF-7/OTV/X-37B as it lifts off the pad. This is the secretive mini-shuttle. The launch was the sixth X-37B mission (OTV-6), U.S. Space Force 7 (formerly known as AFSPC 7), on an Atlas V 501 rocket from Cape Canaveral SLC-41 on May 17, 2020 at 13:14 UTC.

I always welcome your feedback. I can be reached via email at stilley@fit.edu, on LinkedIn at www.linkedin.com/in/drtilley/, and on Facebook at www.facebook.com/stilley.writer. Learn more about my writing at www.amazon.com/author/stilley.

One small step… (NASA)

Information Management in Systems Engineering

Eric Barnhart

....

Abstract

Systems engineering (SE) is arguably not a discipline of design and implementation to the same degree as disciplines such as mechanical, electrical, and software engineering. This chapter posits that systems engineering is a discipline of coordination, communication, and information management. An example of information management for SE is provided and discussed. Greater emphasis on information management has the potential for improving the discipline of SE and increasing the impact of Model-Based systems Engineering (MBSE).

1. Introduction

As systems engineers, we like to think that our jobs are to engineer at a systems level, that is, we "enable the realization of successful systems" as INCOSE would say. In order to do this, we address customer needs, operations, performance, cost and schedule, training and support, and to a lesser degree, test, manufacturing, and disposal. These are all very important areas to address if the systems is to be successful, but they generally fall under different disciplines. If the systems is to meet user needs, an overarching discipline that coordinates and integrates these other disciplines is essential, hence systems engineering.

Systems engineering is arguably not a discipline of design. We don't design integrated circuits or printed circuit boards (PCB). We don't solder components to PCBs. We don't write production C++ or Java code. We don't install blade servers into racks and connect them with a specific network topology. We don't design mechanical parts to be machined on a Computerized Numerical Control (CNC) milling machine. We don't really do the "fun" stuff in engineering.

Instead, systems engineers have their own tasks. We gather the user needs and wants, formalize them into requirements, and create a requirements database. When in doubt about a user need, we perform trade studies to determine the best implementable option, and document that in a trade study report. We define the concepts of operation, systems functions, and logical architecture. We record all that in a set of specifications and architecture documents, or better yet, in an MBSE database. We create mappings between all the entities we defined, and try to document those linkages.

2. Show Me the Data

This is nowhere near a comprehensive list of what systems engineer do, but there should be a pattern emerging. Systems engineers tend to create documents, databases, spreadsheets, relationships and all varieties of information. SE doesn't produce deliverable systems end-products; no software, no hardware, no electronics. Instead, systems engineering produces intermediate products. Systems engineering gathers and synthesizes the information necessary to design and build a product to satisfy the customer. The individual disciplines actually build the product.

Systems engineers answer the critical questions the other disciplines need to know up front to do their jobs right. Systems engineering is a discipline of gathering and synthesizing information, then presenting it in such a way that it can be understood and used for communications both to the development disciplines and also to the customer. Systems engineering is a discipline of information and coordination.

Lately, some educational institutions have issued degrees in systems engineering, but in general we're usually electrical or software engineers who transition into the field. Sometimes we are specialized analysts who can synthesize new technical information. Sadly, very few systems engineers are skilled in information management. For example, how many systems engineers know what a relational database is, or what fourth normal form is? Raise your hands!

3. MBSE as Information Management

I believe Model-Based Systems Engineering (MBSE) is an industry attempt to do more information management for systems engineering. MBSE attempts to capture systems engineering information in a single model and database. Unfortunately, MBSE stops short of fulfilling its promise.

In general, MBSE models will address architectural, behavioral, and structural information, but fail to address other aspects of systems engineering information. Most MBSE tools, especially those based on SysML, cover a limited territory of information concepts. Often, these tools address the end product itself, and not the full field of information concepts that systems engineers deal with. Other tools are used to address other information, such as requirements, planning, testing, trade studies, simulations, and visualization. Although these tools have their own models, seldom are they well integrated.

3.1 The Map and the Territory

There is a very large territory of information that systems engineering covers. When communicating, we do so in the context of the map we've created for that information territory. This can be an issue if somebody in a different context doesn't understand the map we use. If the map omits critical information for the other party, then the communication fails. If the map presents information in an unusable format, communication fails. If the terminology and concepts are incompatible, communication fails. MBSE might provide a limited map that uses common symbols, but much of the remainder of the map exists only in our heads, making it difficult to share. As Korzybski said, "the map is not the territory" [1].

How do you determine what information other people need? How do you anticipate their queries? How do you store and present information in a flexible manner? These are some of the issues that should be addressed to fulfill the information management needs of systems engineering.

In order to better perform our jobs as systems engineers, we need to understand the territory of systems engineering, guide ourselves from point A to point B in that territory, and be able to communicate our results to our stakeholders. We understand our jobs by building and following a metaphorical map of the systems engineering territory. We make decisions about how to get from one point to another. We communicate to our customers by providing them a simplified map of the results that they can understand and relate to.

Why discuss metaphorical maps and not the specific details of the systems engineering process? The answer is simple. Often, we're stuck in a very specific way of thinking. We're locked into specific interpretations of our jobs and processes. By thinking metaphorically, sometimes we can see relationships and be open to ideas we wouldn't see without the metaphorical view.

3.2 What is a Map?

According to Wikipedia, a map is "a symbolic depiction highlighting relationships between elements of some space, such as objects, regions and themes." In a typical roadmap, we see geographic regions and their relationships. Some of the information includes relationships between cities via roads and relative locations. We also see municipal boundaries and geopolitical extents. We can see the relationships between cities and geographic entities, such as rivers, lakes, mountains, and shorelines. Using more detailed views, we can delve down to the relationships between individual addresses and the corresponding cities and roads.

Of course, there are many different kinds of maps. Maps have a specific scope and purpose, a specific theme. Geographical, or topographical, maps are the most commonly used, but there are also many others. For example, aeronautical maps show airspace restrictions, radio frequencies, and general aviation information. Topological maps show relationships and connections, but no extraneous information (such as scale or distance) depending on the application. Network maps show interconnectivity among communications nodes.

In general, maps can do the following:

- Provide location information, to find where you are, whether it be a city address or a network node (navigation).

- Describe the lay of the land.

- Show terrain, entities, and relative placement of the entities whether city addresses or network nodes.

- Show important landmarks.

- Show entities of interest for the subject area≥

- Enable guidance.

- Provide information needed to decide how to get from one node to another.

- Allow you to find the best routes.

There are many things that maps do not do. For example, a map can enable guidance on your available travel options, but the map cannot tell you exactly how to get from point A to point B. Should you travel by car or by airplane? The map cannot answer that. Should you detour to visit an attraction or drive straight through? The map cannot make decisions for you. The map, by itself, cannot tell you the process of getting from A to B, nor the decisions along the way. It only provides the information necessary to enable the decisions. Without the information, you can't make the decisions necessary to follow your process.

To cross terrain, we need both a map and a process for making travel decisions. The common GPS systems in your car or on your phone is a combination of a map and an application to handle decision making. In this case, the decision making is restricted to particular assumptions, such as travel by automobile or the goal is either the shortest distance or quickest route.

If fact, a map comes close to fulfilling the definition of a model:

- It's a visual representation.

- It captures information on structure, data, and inferences.

- It is often formalized, as in the case of GPS systems that store the map info in a database.

3.3 Maps and Processes

System engineering is generally considered to be a set of processes with goals. The main goal is to move from customer concept to design specification that can be handed off to other disciplines, like moving from one map point to another. Where's the information necessary to decide how to move? We describe our processes in Systems Engineering Management Plans (SEMPs), and measure our compliance to our processes, but we very poorly formalize the necessary information. The missing ingredient is the map of the SE landscape that provides the information for the necessary decision-making.

Systems engineers have spent decades developing processes, but far too little time developing and formalizing the map of the SE territory. In order to make decisions, we need a complete model of the systems engineering landscape, including a model of the SE discipline itself, and a model of the customer needs. When building models and maps of decision-making processes, we start to venture into the realm of Enterprise Architecture (EA).

The systems engineering territory does not need to be explored like a poorly known Louisiana Purchase by the Lewis and Clark expedition. We've already covered the SE territory, but we haven't documented it very well. There have been maps of SE, like Figure 1 that depicts the schema from an old SE tool in the 1990s, so the knowledge is there. We just haven't made that knowledge persistent.

3.4 Example: RDD-100 Map

The map shown in Figure 1 is of "Design Guide C" from the tool RDD-100 by Ascent Logic, Inc. Oddly enough, the design guide was only documented in an alphabetically-organized reference guide; this visual diagram was never provided by the vendor. The picture only exists because I spent several days back in the

90's exploring the reference guide to visually uncover these relationships.

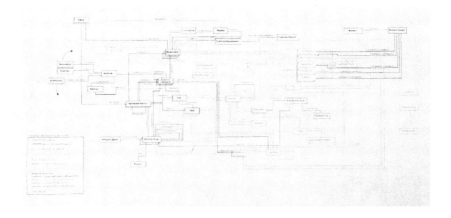

Figure 1: "Design Guide C"

I believe that understanding the basic systems engineering entities and their relationships is the key to improving the entire process of systems engineering, and the solutions we generate.

4. The Simple Requirements Model

What makes a map of systems engineering valuable? How does a systems engineer use a map to get from one place to another in the engineering process? How does a map support decision making?

The previous section addressed what it takes to build a map of the SE territory – an information schema for SE data. I included a real map of a schema created for a tool from the 1990s. That map was far too complicated to explore for an introductory discussion, so I put together a very simple map of systems engineering that I expect many SEs can understand and agree to, at least in principle. The "Simple Requirements Model (SRM)" is shown in Figure2.

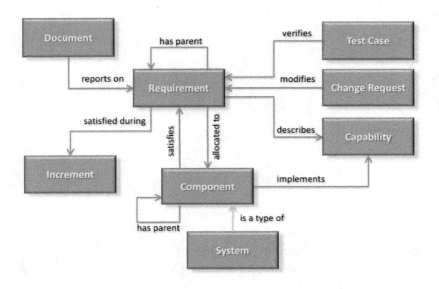

Figure 2: The Simple Requirements Model

The rest of this section explores the SRM to establish some basic concepts about managing maps of information for systems engineering.

4.1 Entities

The model graphically shows entities using blue rectangles. Each entity is a collection of tightly-coupled data that collectively define the entity, and uniquely identify that entity. If you are an aficionado of relational databases, you might think of these entities as representing a schema with tables and relationships. Each individual entity is a row in a table. If you're an advocate of object-oriented methods, you might consider the entities to be objects that realize a particular class. For this high-level discussion, the specific implementation is irrelevant.

Entities also have a set of business rules that are unique to the type of entity. The rules define processes for creating an entity, establishing entity validity, how to perform configuration management for the entity, and others. The business rules are not visible in the entity data; instead, the business rules define the processes and operations behind the entities. Business rules will be unique to the organization that creates the systems engineering map. For this introductory level discussion, the business rules are kept to a minimum.

4.2 Relationships

Entities have relationships between one another, illustrated by lines with arrowheads to indicate directionality of the relationship. Relationships show that entities have a correlation or linkage between them. To capture the meaning of that linkage, every relationship needs at least a name to define it, and rules for how the linkage is made.

Consider your relationship to an automobile. You might be the driver of the car, which is established when you get into the driver's seat, start it, and drive away. You might also be the owner of the car, which is established by the title to the car and kept by a local government entity. You and the car are two very different entities and have two (or more) very different relationships.

4.3 Exploring the Map

As described above, a well-constructed map of an information territory, or domain, represents well-defined entities with not only a name and perhaps a descriptive definition, but also with a set of attributes necessary to characterize the entity in the particular context. Each of the elements in the SRM needs its own set of attributes – including the entity at the center of the SRM: the requirement.

According to ISO/IEC/IEEE Standard 29148-2018 [2], a requirement is:

1. A condition or capability needed by a user to solve a problem or achieve an objective.

2. A condition or capability that must be met or possessed by a systems or systems component to satisfy a contract, standard, specification, or other formally imposed documents.

3. A documented representation of a condition or capability as in (1) or (2).

For our purposes, a requirement is most closely like definition (3), the documented representation of (1) and (2). Since it is a defined element of this model, we shall refer to the entity as a «requirement», where the guillemets («») act as a flag to indicate that we are talking about a well-defined element of the SRM.

If you choose to use an indicator of a defined entity in your documentation, you can choose any indicator you like, or no indicator at all. I advocate flagging any reference that has a well-defined meaning to let readers know they're looking at something using a precise definition, and not a random or ambiguous term.

Not only does the «requirement» have a definition, it is also characterized by a specific set of attributes that describe the «requirement» and only the «requirement». The attributes of a «requirement» in the SRM are shown in Figure 3.

Requirement Attributes

Attribute	Meaning	Necessity	Type
ReqID	Short unique identifier for the requirement	Mandatory	(KEY)
Text	The full text statement of the requirement	Mandatory	Text
Guidance	Formal explanatory material that further guides or constrains the requirement; this material is controlled and tested to the same degree as the requirement text.	Recommended	Text
Rationale	Explanation of why this a valid requirement (i.e. why this requirement exists)	Recommended	Text
CM State	Current state of configuration management for this requirement	Recommended	Enumerated
Notes	Informal comments on the requirement; not controlled or tested.	Optional	Text
Category	Company process unique category	Optional	Enumerated
Abbreviated Text	An abbreviated description for convenience	Optional	Text
Security	Flag to indicate this requirement is a security issue	Optional	Boolean
Safety	Flag to indicate this requirement is a safety issue	Optional	Boolean
Legal	Flag to indicate this requirement is a legal or regulatory issue	Optional	Boolean
Rule Check	Flag to indicate this requirement passes company internal process checks	Optional	Boolean
Author	Name of original author	Recommended	Text
Origination Date	Date of first creation	Recommended	Date
Change Date	Date of most recent update via change request	Recommended	Date

Figure 3: The attributes of a «requirement» in the SRM

In the context of the SRM, a «requirement» can be characterized with only two required attributes: a unique ID and the text of the user need. (SysML defines exactly these two.) In the SRM, there are 15 requirement attributes defined, and only two are mandatory.

Since the «requirement» entity is at the center of the SRM domain, the next step will be to see how it relates to the other entities in the SRM.

4.3.1 «Requirement» «has parent» «Requirement»

A set of «requirement»s tends to have a hierarchical organization, so a «requirement» can have a relationship to another «requirement» called its parent, or «has parent». The relationship is shown with a directed arrow. The hierarchy can be discerned by recursively following «has parent» relationships until we find a «requirement» that has no parent «requirement»s.

In a more complex model, the «has parent» relationship might be replaced or complemented with more detailed relationships, such as «decomposes», «disambiguates», or «derived from». The SRM uses the simpler «has parent» relationship.

4.3.2 «Document» «reports on» «Requirement»

A hierarchical set of requirements is usually represented not in a simple list, but in a document or specification. A «document» is a separate entity from a requirement because a document has a different set of characteristics. Documents will have a formal ID, title, author, revision, multiple sections and subsections and release sensitivity. None of these apply to individual requirements. It makes sense that a «document» should be stored as a separate entity from a requirement.

This is significantly different from legacy DOORS implementations, in which requirements and documents are misguidedly mixed together into a confusing morass. (It is possible to separate the two using DOORS 9.x, but that requires the design of the database to differ from the originally intended design of the DOORS application.)

Furthermore, «document»s differ from «requirement»s because «document»s have different methods to operate on and maintain them. Instead, «document» «report on» «requirement»s, or inversely, «requirement»s are «reported on» by «document»s.

Publishing a specification is a process of selecting the «requirement»s from a model based on a query of the requirement allocation to a component, and then building a «document» to organize and report on the selected «requirement»s.

4.3.3 «Requirement» «allocated to» / «satisfies» «Component»

A «requirement» states what the customer needs, but that need is delivered as a «component» of the overall deliverable system. This is how the SRM ties together the descriptive model of «requirement»s with the structural architecture model of «component»s.

A typical business rule maintains that a «requirement» is allocated to exactly one «component». (If it were allocated to more than one, then the «requirement» would be stating more than one goal and should be decomposed into multiple requirements.) The inverse relationship says that a «component» satisfies a «requirement». A «component» can satisfy multiple «requirement»s.

A «component» is a separate entity from a «requirement» because it has unique characteristics that describe it as a physical entity. «Component»s will have a unique identifier, a set of functions, a set of constraints, physical characteristics that describe it as a controllable physical item, procurement attributes and others. In addition, «component»s are managed and controlled as configuration items, not as «requirement»s.

4.3.4 «Component» «has parent» «Component»

The architecture of a systems generally is a hierarchical structure of «component»s. Similar to a «requirement», a «component» has a relationship to a parent «component» via the «has-parent» relationship. Traversing the «has-parent» relationship reveals the structural hierarchy.

4.3.5 «System» «is a type of» «Component»

In the SRM, «requirement»s are «allocated to» «component»s, which allows some flexibility in how a business defines a «component». Since this is a systems engineering model, the model needs to have a «system» in it. The SRM shows that a «system» is a specific kind of «component» using the relationship «is a type of». In a larger and more useful SE model, the «system» and «component» would have a much more detailed model surrounding them and describing more relationships.

4.3.6 «Requirement» «satisfied during» «Increment»

In many projects, the development effort will be spread over several phases, or implemented in «increments». A business rule might state that a «requirement» will be «satisfied during» exactly one increment. (If it were «satisfied during» more than one, then the «requirement» would be stating more than one goal and should be decomposed into multiple requirements.)

4.3.7 «Test Case» «verifies» «Requirement»

In order to prove that a «requirement» is met, the «requirement» needs to be tested and verified. A «test case» defines the testing and verification approach and is related to multiple «requirement»s through the «verifies» relationship. A «test case» is a unique entity from a «requirement» not only because it has unique descriptions and goals, but is finalized during a later lifecycle phase of the program, generally after the requirements set has been baselined.

4.3.8 «Change Request» «modifies» «Requirement»

In order to control changes to a requirement set in an orderly fashion, each «requirement» is generally placed under control of configuration management. A business rule might state that a «change request» is necessary before modifying a «requirement». The «change request» might have attributes that look a lot like the attributes of a «requirement», as well as additional attributes for status and tracking of the request. The «change request» is handled differently from a «requirement» and lacks the other relations of a «requirement». Given these factors, the «change request» is an independent entity in the SRM.

4.3.9 «Requirement» «describes» «Capability»

A «requirement» states what the customer needs, but ultimately the customer has a specific desired effect the customer wants to see in the end product. That desired effect is captured by a «capability» entity. Typically, multiple «requirement»s will «describe» a «capability». This is how the SRM ties together the descriptive model of requirements with the results-driven behavioral model of «capability»s.

4.3.10 «Component» «implements» «Capability»

Finally, the «component»s of the end product must exhibit well-defined, desired behavior to satisfy the customer. The SRM models this by showing that a «component» «implements» one or more «capability»s.

A business rule might allow a «component» to «implement» many «capability»s but insists that a «capability» be implemented by exactly one «component». If a «capability» were implemented by multiple «component»s, then the «capability» would need to be decomposed into multiple subcapabilities. The SRM, being simple, does not support this.

5. Summary

The Simple Requirements Model presented here is so simple the reader no doubt found multiple deficiencies and inconsistencies between the SRM and their own company's systems engineering approach. Still, creating and explaining the SRM proved to be a surprisingly tedious process. Part of what makes it so tedious is that we feel we already know these entities and relationships. If we knew them so well, they would be written down, wouldn't they? If everybody knew them by the same exact definition, wouldn't the systems engineering process go much faster and be more accurate? Despite feeling that we know them, very few organizations seem to have really taken the time to explicitly define the entities and relationships. As a result, when moving from company to company, and even project to project, we waste a lot of time reinterpreting our map for the SE process, and stumbling over minor misconceptions and terminology.

When looked at from a different perspective, systems engineering is a discipline of coordination, communication, and information management. In order to coordinate and communicate, systems engineers need to speak a well-defined language and follow a well-plotted map of the systems engineering territory. Greater emphasis on information management has the potential for improving the discipline of systems engineering and increasing the impact of MBSE because better maps of meaning and relationships will enable higher quality communications and coordination.

References

[1] Korzybski, A. "A Non-Aristotelian System and its Necessity for Rigour in Mathematics and Physics." *Meeting of the American Association for the Advancement of Science*, New Orleans, Louisiana, December 28, 1931.

[2] ISO/IEC/IEEE International Standard - Systems and Software Engineering - Life Cycle Processes - Requirements Engineering (29148-2018). Published Nov. 30, 2018. Online at https://standards.ieee.org/standard/29148-2018.html

About the Author

Eric Barnhart graduated with a degree in Electrical Engineering, but found himself thrust into the world of software with his first job at the NASA Johnson Space Center. While there, he became exposed to the real world of engineering, including bureaucracy, bloated processes, and failed programs. Realizing that the lack of systems thinking was the number-one problem behind most engineering project failures, he gravitated to systems engineering. After 30 years in technology and engineering, and 20+ years as a systems engineer, he has accumulated a wealth of lessons learned that he shares at his website, https://vmcse.com. Contact him at Eric.B@vmcse.com.

# # #

The historic Cocoa Beach Pier.

(Zhukova Valentyna; Shutterstock 715664839)

The ESNOS Spacesuit

Tiziano Bernard and Lucas Stephane

....

Abstract

When thinking of systems engineering, one thinks of requirements, project management, interdisciplinary efforts, hundreds of meetings, and overall a highly organized, lengthy approach to project execution. The story of how ESNOS - the Enhanced Space Navigation and Orientation Suit - prototype came to be is nothing of the sorts. The concept was born from a random inspiration, and the systems was constructed within a week. As much as ESNOS could be considered nothing more than an academic exercise, the space suit was created according to human-centered design thinking to be applied onto the astronautical domain. ESNOS is a great example of what Florida's Space Coast is all about: a region where scientists exchange ideas that literally suck the air out of a room. And then, just like we did, they build those ideas and branch new projects from it.

1. Introduction

In October 2015, I (Tiziano Bernard) was walking down the main staircase of the Olin Engineering Complex at the Florida Institute of Technology. It's one of the most modern-looking buildings on campus and I used to consider it my second home. I completed a Bachelor's degree in Aerospace Engineering and was now pursuing a Master's degree in Flight Test Engineering.

Most of my professors and labs were located in the Olin building. Moreover, as an instructor for some of those labs, I had an office on the second floor.

It was routine for me to leave the office at night and walk down the main staircase. That night, however, my eye caught the aerospike on display as I walked onto the staircase. In aerospace propulsion, an aerospike is simply put a "reversed rocket nozzle" that maintains efficiency with changing altitude. What really caught my attention was that external flow around the "spike." I'm not sure how, but I pictured a person instead of the spike: I imagined a person being surrounded by a device that could transmit sensations. A suit, perhaps.

As a pilot and flight test engineer in training, I became acquainted with military personnel and thought of one of them in a fighter aircraft with a suit that would aid them in the identification of other traffic, unidentified objects, and so on. I imagined a suit that was so much more than a wearable, making the surroundings tangible to the user. I was not exactly fluent in human-systems integration or user experience, and only confided my idea to my "advanced interaction media" professor, Dr. Lucas Stephane. It was simply an idea. And it remained so for a whole year.

1.1 The Human-Centered Design Institute

Exactly a year later, in October 2016, I began my doctorate in Human-Centered Design at what is known as the Human-Centered Design Institute (HCDi). To date, I find it difficult to define exactly what it comprises. By late 2018, I was a mixture of one's academic background (aerospace, in my case), systems engineering, human-systems integration, and a touch of human factors. My dissertation, for example, was an application of cognitive engineering for flight test methods using virtual reality technologies. A mixture of disciplines, indeed.

In my experience, one of the most challenging points of getting a Ph.D. is narrowing down what will essentially become your expertise, the heart of your dissertation. In that great search for a topic, I thought of that suit concept. Could I design and create something that makes external elements tangible to the user, effectively become embodied?

My advisor, who became in fact Lucas Stephane, believed in the idea and recommended I express my thoughts to an expert, someone who had extraordinary experience in military applications and also had a vision for research interests.

1.2 Meeting with an Astronaut

On Florida's Space Coast, an astronaut is not extremely difficult to find, however having a meeting with one of them is definitely not to be considered a simple event. Florida Tech has various resident astronauts, including moonwalker Buzz Aldrin. The astronaut that I spoke to was Winston Scott, an experienced Navy Captain, naval aviator, and Space Shuttle veteran. He was one of the figures who encouraged me to become a pilot and served as a member of my Master's committee in flight testing. He was the Senior Vice President for External Relations and Financial Development and now is the Senior Advisor to the President. He later also joined my doctoral committee.

I always found talking to Capt. Scott quite therapeutic. His military preciseness and astronautical transparency always filtered my thoughts into linearity. When I showed him a brief slideshow on my computer outlining the suit concept, he had plenty to discuss. It appeared that a variation of my idea was already in existence, and that of course (I sometimes forget that I'm talking to an experimental test pilot and astronaut!), he had tested it. He described a vest that provided haptic feedback (one of the technologies that I was looking at for my suit) for military pilots [1][2]. During a series of experiments in Pensacola, he was able to fly a Navy T-34 simulator configured for instrument flight using haptic feedback as a means for navigation. As a NASA astronaut, he was evaluating possible applications for extravehicular activities (EVA) to avoid disorientation. He recommended that I look into other applications, including automotive, aeronautics (with particular emphasis on helicopter hovering guidance), and those EVA astronautical applications.

Although thrilled, the possibility of transforming this idea into my doctoral project got more distant, as the contribution in the field would have been difficult to defend given the established efforts of the US military. I continued, however, to think of the suit and its power of embodying external inputs. Wouldn't it be incredible to build one, test the usability, and discover new applications?

Just a week later, the opportunity showed up.

1.3 The Thales Arduino Challenge

The city of Melbourne, Florida, is one of the fastest growing industry centers in the United States. Home to the Florida Tech research park, it includes engineering excellences such as L3Harris Corporation, Northrop Grumman, Lockheed Martin, Rockwell Collins, Embraer Executive Jets, Thales Avionics, and many others.

HCDi, for which I was a doctoral research assistant, is one of many university entities that thrive in the interplay of government-, academic-, and industry-created challenges. Dr. Stephane was the one who stopped me in the hallway and told me about a new opportunity: The "Thales Arduino Challenge," which was organized by the avionics company to allow students to express ideas by providing free equipment in the form of Arduinos. It fosters creativity and ingenuity, since Arduinos can be used in any field and at any scale, making the prototypes affordable and relatively easy to build.

As with many HDCi projects are, the suit concept required multidisciplinary efforts. An expertise in software was undoubtedly necessary, as all things in today's world are somehow run by a computer. The architecture of the suit, which required wiring and a control system, called for electrical engineers, and finally, the operational design and overall systems architecture required both aerospace and human-centered knowledge. I was confident of being able to manage the latter two, having done my undergraduate studies in aeronautics and astronautics and currently pursuing a field which included systems engineering and human-systems integration. The rest of the expertise came from the project advisor, Dr. Stephane, and three of his students: Kushal Vangara was a software engineering graduate student focusing in cybersecurity, and Andrea and Vincenzo Miale were two incredible undergraduate electrical engineering students. In particular, they had excellent knowledge of smart technologies, especially conductive inks. The team was indeed strong and highly motivated. I find motivation pivotal in generating ideas.

With regards to creativity, I found that a requirement is the absence of materialism. I believe that, given unlimited finances and a certain detachment from material goods, creativity blooms. By this, I mean the ability to not become attached to things. For example, if I become attached to a certain camera, maybe because I spent all my savings on it, it will limit what I will personally allow myself to do with it. Since the financing came from Thales and some additions by Dr. Stephane, there was very little attachment to the devices, allowing for the most creative of ideas to be developed.

"Maritime? Transportation? Shall we stay in aerospace?" There were so many concepts and ideas flying around the HCDi conference rooms. It was decided, however, that a focus in aerospace would be more aligned with our knowledge and the interests of Thales. "Aeronautics or Astronautics?" was the next question. Needless to say, the multiple of ideas and concepts ended up pointing at a haptic feedback suit. Since the US Navy had already developed the technologies for pilots and naval aviators, we focused on astronautics: making the outside surroundings tangible by providing navigation and orientation feedback. Thus, the "Enhanced Space Navigation and Orientation Suit – ESNOS," was born [3].

2. Theory

The need for an astronaut assistance wearable could be a logical step in the development of devices for planetary colonization and extravehicular activities. Although current EVA operations are still considered an uncommon event, future EVA operations will become daily and routine. Therefore, it was thought that haptic feedback could allow an improved interaction with the environment. Stimulating the sense of touch could allow both navigational cues (i.e., provide a direction to follow) and allow for orientation assistance (i.e., aid in the perception of "up," "down," and so on).

All of these improved interactions can be summarized as "embodied situation awareness." Typically, this awareness is increased by inputs coming from the outside world: an external reality dominated by an environment and a sociotechnical system (collaboration between people and technologies).

To better organize our thoughts, the team turned to Dr. Stephane. He often led these discussions in what we officially called "knowledge elicitation" sessions. We do however, jokingly, call them "knowledge extraction" sessions, as he is able to take our chaotic thoughts and brilliantly align them on a whiteboard (as shown in Figure 1). His experimental psychology background was a definite benefit in this regard.

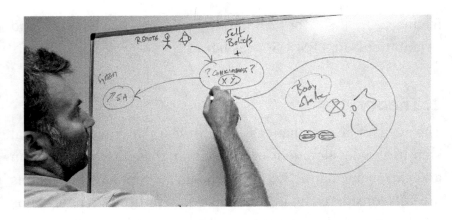

Figure 1: Lucas Stephane at the board, graphically describing the interaction between ESNOS and consciousness, self-beliefs, situation awareness, and body state.

The suit is equipped with five vibration motors: two in the front, one on each arm, and one in the back. This allows directional guidance in the horizontal plane. For proof-of-concept purposes, we only focused on one axis.

A control panel, located in the back of the suit, hosts the Arduino and the battery to power the system. The controls for the suit motors are located on a separate control station on a laptop. The communication occurs via Bluetooth technology protocols. The laptop has an interface that allows the user to control each motor, activate a "panic" mode which activates all motors at once (also controllable through a "conductive ink" switch on the suit's front). The schematic, as created on the whiteboard during a meeting, is shown in Figure 2.

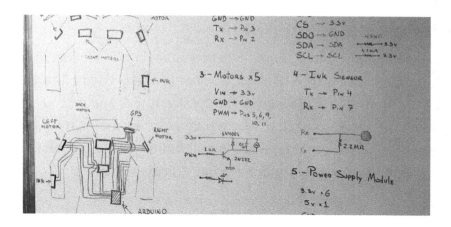

Figure 2: ESNOS schematic. Whiteboard drawings were pivotal in both presenting concepts and keeping up to date with modifications and improvements.

The control panel also provides data on accelerations (through a 3-axis accelerometer) and GPS location (through a GPS module). The user interface is shown in Figure 3.

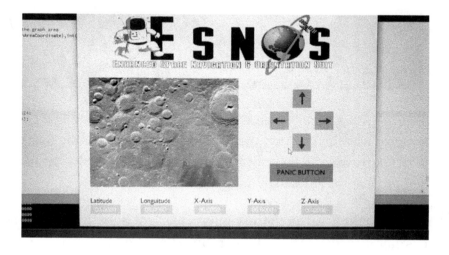

Figure 3: The ESNOS user interface. It provides accelerations, GPS position and a map display, the panic button activation, and the four motor commands.

3. Practice

I was lying on the floor outside the Harris Florida Tech Commons building where HCDi was located. I was, literally, laying on the sidewalk, staring at the sky. It was about 1:30 a.m., and I took a small break to take a picture of the moon. It was the "super moon" I was interested in. I definitely could not miss the opportunity to take a picture from Florida's Space Coast that night!

Figure 4: The Super Moon was spectacular. What truly made it magical was the fact that the picture was taken from the Space Coast. Less than an hour away is the Kennedy Space Center and our project was advised by an astronaut.

As I was looking through the camera, a bright light disturbed my peripheral vision. As I tilted my head to the side, I realized I caught the attention of a campus security officer. I reached for my wallet, pulled out my Florida Tech ID card and waived it at the large security vehicle which stopped beside me. The officer wasn't too worried about my presence. Rather, he wanted to see the pictures of the moon I took! Finished with security, I figured it was time to go back to work.

Upstairs, on the third floor meeting room of HCDi, the wiring, computers, and astronaut suit were all over the place. The team was ferociously typing, connecting, taping, and discussing. It was that night that ESNOS became reality.

Figure 5: The ESNOS student team. Many meetings occurred throughout the project. One of the last ones was about the test plan for usability purposes.

Once completed, the team decided to proceed with usability tests. With no micro-G environment available, we tested the systems by having Andrea wear the suit, get blindfolded, and through the vibrations see if she could navigate around a maze. The location was pretty easy: the hallway on the third floor of our building. We strategically borrowed rectangular plant vases from the International Student Office and built a path in the hallway. To our great surprise, the suit successfully allowed Andrea to reach the end of the maze.

What was truly remarkable was that it all happened, from design to testing, in a single week.

4. Lessons Learned

Throughout the project, we were able to exercise many theories and determine their effectiveness. Although these conclusions are based on just one project, the lessons learned seem valuable.

4. 1 Agile Project Management Works

Many companies today live in what is referred to as "ordered chaos." A hierarchy is still undoubtedly present in the overall picture, but in small working groups the "all hands-on deck" approach is appropriate. Originally, there was a thought on micromanagement. Interfering is not micromanagement if the intent is collaboration.

Let's express this in practical terms. When deciding the wiring of the suit, I personally did not interfere, since my expertise is not in wiring. I was, however, involved when the wiring had to be aligned with the operation of the suit (e.g., will the wiring impede arm movement?). The clear distinction of expertise made the management of the project more transversal, allowing a more peaceful collaboration and overall effectiveness in creativity.

4.2 Crazy Can Be Good

I recall Dr. Stephane recommending the use of a laser scanner as extra equipment on the suit. I shot down the idea as being an overkill. Looking back, he was right. Crazy ideas can suck the air out of the room and Lucas often has those. A big lesson is therefore to always consider all ideas, no matter how disproportionate they might appear.

4.3 Interdisciplinary Efforts Are Key

No big project can be done with a one-man team. A successful subdivision of expertise and tasks is key in optimal systems integration. The team performed with very little overlap between members. In this way everyone was able to work independently towards the same goal.

4.4 Team Spirit Determines the Final Product

The way the team is involved in the project undoubtedly influences its outcome. The project was part of a competition and this increased the motivation of the team members. Moreover, the limited cost of the systems truly fostered creativity.

5. Summary

The ESNOS team was formed as part of an academic challenge set by Thales using only an Arduino Uno and a set of limited-cost devices. The team was allowed two weeks to work on it: one week spent on discussing what to design, and a second week to design, build, and test the system.

Through agile project management, human-centered design thinking, and great human-systems engineering, ESNOS became a reality. Moreover, the project was elected "best project at Florida Tech" by Thales.

A former professor once said that a great project is one that opens a new field of study. I am proud to say that one year and half later, ESNOS became the skeleton from which a new doctoral student is designing a sensory enhanced suit for astronautical operations [4].

It might sound cliché, but it is true that working on Florida's Space Coast has a component of elegant magic. I took the liberty of writing this text in an informal manner – something I'm not very used to as an academic. I found it very relaxing, and I hope that all the emotions, sensations, and "awes" of amazement are perceivable by the reader.

References

[1] D. Myers, T. Gowen, Angus Rupert, B. Lawson, J. Dailey. 2015. "Coalition Warfare Program Tactile Situation Awareness systems for Aviation Applications: Simulator Flight Test." United States Army Aeromedical Research Laboratory, Aircrew Health and Performance Division Report 2016-07.

[2] Angus Rupert, B. Lawson, J. Basso. 2016. "Tactile Situation Awareness System: Recent Developments for Aviation." *Proceedings of the Human Factors and Ergonomics Society Annual Meeting* pp. 721-725. Washington, D.C.

[3] Tiziano Bernard, Andrea Gonzalez, Vincenzo Miale, Kushal Vangara, Winston E. Scott, Lucas Stephane. 2017. "Haptic Feedback Astronaut suit for Mitigating Extra-Vehicular Activity Spatial Disorientation." *Proceedings of the 2017 AIAA SPACE and*

Astronautics Forum and Exposition. Orlando, Florida. https://doi.org/10.2514/6.2017-5113.

[4] Brandon Cuffie, Tiziano Bernard, Yash Mehta, Mehmet Kaya, Winston E. Scott, Lucas Stephane. "Proposed Architecture of a Sensory Enhanced Suit for Space Applications." *Proceedings of the 2017 AIAA SPACE and Astronautics Forum and Exposition.* Orlando, Florida. https://doi.org/10.2514/6.2018-5153.

About the Authors

Tiziano Bernard is an aerospace engineer and pilot, currently designing advanced flight decks at Garmin International. He also writes for the Italian national newspaper *ilGiornale*, covering science, aerospace, and Italian-American relations. In 2019, he was elected to FORBES Italy's 30 under 30 list for science. He has degrees in flight test engineering and aerospace engineering, and holds a Ph.D. in human-centered design from the Florida Institute of Technology (2018). Contact him at bernard.tiziano@mac.com.

***

Lucas Stephane is a Senior Scientist AR/VR with the Institute for Energy Technology in Norway. He was previously an Assistant Professor of Human-Centered. He has over 20 years of experience in industry and research institutes as an expert in cognitive engineering and human-systems integration. Holder of two patents with Airbus, he is a member of the National Academy of Inventors (USA). His experience is interdisciplinary, focusing in aerospace, automotive, serious games, and virtual, augmented, and mixed reality technologies. He holds a Ph.D. in human-centered design from the Florida Institute of Technology (2013). Contact him at acerodon@gmail.com.

# # #

Astronaut Buzz Aldrin, now a resident of Brevard County, on the moon during the Apollo 11 moonwalk in 1969. (NASA)

How to Succeed at Driving Change

Joseph Bradley and Kaitlynn Castelle

....

Abstract

Driving change in an organization involves drawing a picture of the desired future, focusing on Why, What, When, How, Where and Who. Why is probably the most important question, since it provides the motivation for What and How; without understanding Why, your effort is likely going to miss the biggest reasons for driving change. In most cases, hold off on answering the How for a while; addressing it too quickly might keep you from figuring out the What. Answers to the rest of Kipling's six honest serving men can be discovered by following a seven-step process that explores the capabilities allowing an organization to address complex problems.

1. Introduction

> *"If you want to succeed at driving change, practice drawing pictures. Specifically, practice drawing pictures of either what the future looks like or what the journey to the future looks like."* April Mills (2011)

I often run workshops helping organizations drive change. This involves identifying the organizational capabilities for addressing complex problems. As shown in Figure 1, a seven-step process can be used to identify these capabilities. The process allows us to explore the Why, What, When, How, Where and Who of the desired change.

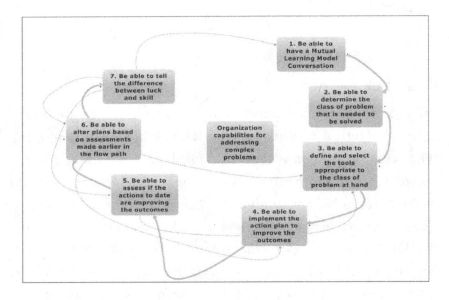

Figure 1: A process for addressing complex problems

Note that we do not attempt to solve these complex problems, since by their very nature they are extremely hard to solve. Partly, that is because a problem is complex if there are many different stakeholders and their solutions are mutually incompatible, their perspectives and needs will vary, thus their proposed solutions may be mutually incompatible, particularly in the case of perceived resource scarcity, which is almost always the case. Instead, the goal to make things discernably better than they are today.

2. Step 1: Improved Conversations

In the first step of the process, I start to discuss the question of "What is needed?" and there are lots of ideas from people that I ask. That is good, but as usual, getting people to agree is pretty hard. We need to be able to converse about the complex problem without getting locked into our mental models and typical patterns of thinking that will constrain any type of solution we try to produce.

Chris Argyris and Daniel Schön postulated two kinds of conversations: (1) Unilateral Control Model, and (2) Mutual Learning Model. In the former, the climate allows questions, unless they reinforce the "correct" answer. Hard topics are avoided, decisions are often made about people without their presence or input, and problems often appear as surprises, though many people already knew about them.

In organizations that permit or encourage the Mutual Learning Model conversation, people can ask questions without being shut down. Proposals are subject to debate, rejection, or modification, and not treated as final decisions masquerading as proposals for discussions. In this type of climate, personnel decisions are made collaboratively, diplomatically, and transparently. This is what we want.

You might say, "But I have lots of conversations!" And you probably do. But step back and think about the conversations you've had lately, whether one-on-one, in a meeting, or in a setting where someone is addressing a large group. How did those conversations go for you? Was there wide agreement? Did the actions that were agreed upon happen, when and as they were promised? Were realistic action items even proposed?

In this reflection, you may have realized that a lot of conversations were not fruitful, in which case, you have merely joined most of the world. Only a small fraction of the world has fruitful conversations. If they are fruitless, aggravating, and frustrating, then *learn how to have a better conversation*. In doing so, we learn that our suspected problem appears to be much more complex than we initially realized.

3. Step 2: Problem Determination

Step 2 of the process begins when we can converse productively, and we need to decide what tool(s) to use to work on the problem. It does little good to use a hammer to cut a plank of wood. We need to be able to divide our problems in classes, and then choose the correct tools to match the class of problem. We will be a lot better off when we are able to have this conversation, as it will help us identify the correct problem class and thus the correct set of available tools. This is inherent to being an engineer. If a new solution is created for every problem, you are dealing with a scientist, not an engineer.

4: Step 3: Tool Selection

Once the class of problem has been determined, selecting the appropriate tool(s) is the next step in the process. If we have trouble with the conversation, we can fall into any number of traps, including a black hole as we disagree over the tools and never get started, or we chose the tool the loudest person likes and knows, and spend too much time talking about the problem but not making progress on the problem. You are now in the How question phase: How can we make this tool work to address our identified problem?

You also are in the Who question phase. You need people to make this happen. Let us assume you had a good conversation, and thus could decide on the correct tool(s), How to use the tool, and Who to get on the team. What comes next?

5. Step 4: Plan Implementation

Step 4 of the process is plan implementation. You need the ability to execute the project plan to improve the situation using the selected tools. *This means you need people*, with the skills to execute, and the hardware and the software resources they will use to get done whatever you decided to get done.

6. Step 5: Action Assessment

Step 5 of the process is action assessment. Once you get moving, you need to be able tell if you are making a difference, moving in the right direction. Hopefully, you figured out the way to decide that already, but if not, join the rest of us who try to figure it out on the fly. But, you do need a way to know if things are getting better.

It can be pretty easy to fool yourself, so it helps to have hard numbers, a measurement or yardstick that is not relativistic. What does relativistic mean? Lots of people adjust their measurement device to make poor results looks better than they are. For example, "Oh, we only got a 3% improvement when we were aiming for 30%, so we redefined our measurement and we are now EXCELLENT." The scale was relativized.

7. Step 6: Plan Alteration

Step 6 is plan alteration, which is a challenging activity. Assuming you have resisted relativistic effects, so your measurements are good, but the effort is not going anywhere; nothing is getting better. What should you do?

I propose that you need to be able to alter course, starting with asking some questions: Did the world change? Do we have the right people? Are we allowing them enough time to work on this project? Has our sponsor delivered the support we need? Those are just a few of the questions one might ask to assess Why your effort has stalled.

Alternatively, you may have executed well and things are going smoothly. You still need to assess your progress, and make this a regular routine for your team to reflect on overall progress. Ask the following questions:

- Were your initial goals too low?

- Can you go further than you designed? (Should you consider "requisite parsimony"?)

- Have conditions changed and you might be headed for a stall?

- Are you losing support of a sponsor?

- Are better tools available?

- Is there a way we can improve our current work processes?

The point is – good or bad – you need to periodically assess the identified plan, and how it has been and is currently being executed: What and How the effort to attain it is proceeding, and be able to alter course based on that assessment. Through this intentional effort, we have a self-correcting process that is customizable and flexible for future technical and program changes.

Often, your leadership and your team are committed to the current plan, no matter how you are doing. It can be tough to alter course, but you have the measurements to back up your analysis. Convince them that it is okay to adapt to the situation as it changes, or as we learn more. Teach them to expect and welcome new knowledge about our problems, and with that new knowledge will mean that our plans must adapt as well.

8. Step 7: Project Assessment

The final step in the process is project assessment. Eventually, you will call the project complete. Life is better, or it isn't, but the end is nigh. A crucial steps needs to occur. The assessment needs to include the analysis of the following question: *Was this luck or skill?* It is crucial to answer this question. Everyone who touched this project will get the credit – or the blame.

Think of all the one-hit wonders you have seen. Was that result from luck or skill? You really need to decide which are luck and which are skill, otherwise you are playing roulette on every project. If you promote people based on luck and they get a bigger project the next time, your risk is now raised and you may not even know it – until it's too late.

9. Summary

The red lines in Figure 1 represent feedback from some of the boxes to earlier boxes. They are arranged in a loop. This is meant to convey that this is a continuous process, not a use-once-and-forget process.

If we go back to April's quote from Section 1, she advocated drawing either a picture of what the future looks like, or a map to get to the future. Figure 1 is actually both. It shows a future where your organization has the capabilities to work on complex problems, it tells you what those capabilities are, and even though we did not discuss the feedback, it shows how the feedback occurs to improve the overall process and inform specific capabilities.

The map also shows you how to get to the future: you decide you want or need this capability, and then you start building each of those that you do not have today. You might build linearly, but since it is a loop, you don't have to be trapped by linear logic. You can develop all along the loop using the same feedback capability to determine where you are in getting to having the capability to work on complex problems.

About the Authors

Joseph Bradley is president of Leading Change, LLC. He is also a retired US Naval Engineering Duty Officer with extensive experience in the operation and maintenance of ships and shipyards. His most recent project is helping to create a sustainment system for the future COLUMBIA class submarine.. He has been published in peer reviewed journals and spoken internationally on the topics of complex system governance, competency models, and IT frameworks. He holds a Ph.D. in Engineering Management and Systems Engineering from Old Dominion University (2014). Contact him at josephbradley@leading-change.org.

***

Kaitlynn Castelle is a Product Support Strategist for Amentum. She is also an Adjunct Assistant Professor at Old Dominion University in the Engineering Management and Systems Engineering Department, where she teaches project management, agile development, and provides expertise on improving organizational agility and transforming government programs. She holds a Ph.D. in Engineering Management and Systems Engineering from Old Dominion University (2017). Contact her at Kaitlynn.Castelle@amentum.com.

# # #

The Brevard Symphony Orchestra performing in the King Center at Eastern Florida State College. The King Center is the main venue for artistic events of all types on the Space Coast. (Scott Tilley)

Technology Qualification

David N. Card and Michelle E. Novaes-Card

....

Abstract

Innovation has become a principal business concern. Product innovations are often incremental – new features added to established capabilities. This is especially true in the realm of the Internet of Things (IoT). This chapter introduces the general process and key concepts of technology qualification (TQ) as a technique for establishing the quality, safety, and security of complex systems involving components with varying degrees of innovation. TQ is used extensively for hardware in the maritime and offshore industry. Software is increasingly important to complex and even everyday products. While reused software often is assumed to be safe and reliable, use in a new application can prove problematic. TQ promotes the adoption of new technology by giving early adopters confidence in it, including software. This chapter is based on experience in the maritime industry, but the lessons are broadly applicable.

1. Introduction

Effective and early adoption of new technology has become a critical factor in business success. Today's new technologies are largely software dependent. In order to explore this situation, it is instructive to examine the motivation of early adopters and the nature of the quality challenges they face.

1.1 Motivation of Early Adopters

History shows that new technologies are adopted gradually over time, as illustrated by the Everett-Rogers technology adoption lifecycle model shown in Figure 1 [1]. Hall and Khan [2] argue that the technology adoption decision is a cost/benefit trade-off. If such analysis were easy and sufficient, technology adoption would progress much faster than it has historically.

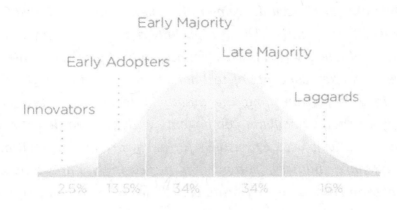

INNOVATION ADOPTION LIFECYCLE

Figure 1: Technology adoption lifecycle model (1962)

However, in many industries, the risk posed by new technology must also be considered. The consequences of unacceptable performance may be much greater than any potential benefits. Also, the nature of new technology is that costs, benefits, and risks are uncertain. Technology Qualification (TQ) offers an approach to reducing the risk associated with new technology as well as gathering cost and benefit data, and thus facilitating its adoption.

Most organizations wait for conclusive evidence that the technology is effective before trying it out. However, early adopters do not blindly ignore the risks of innovation, but make a decision based on the limited information available. When something is new, typically there isn't much information about its performance, so making reliable comparisons with existing technologies may be difficult. TQ helps address these problems.

Candidate adopters often seek quantitative evidence that the innovation is more efficient, reliable, or cheaper than existing options. Because established technology is accepted as safe and efficacious, its performance often is not measured, so even if the innovator conducts a carefully instrumented case study, there may be no baseline to compare it against.

For example, several medical diagnostic support systems have been developed in recent years, but none is in widespread use [3]. Frequently cited adoption obstacles include lack of user friendliness and lack of confidence in the accuracy of the systems diagnoses. These issues may be summarized as concern for quality, especially in regards to the system's software.

It is difficult for a non-specialist to evaluate the degree to which a complex systems such as this has been verified and/or qualified for use. TQ provides a systematic and reproduceable method of qualification relative to product objectives. By establishing that an innovation is safe, reliable, and fit for use, the adopter can be assured that the innovation, at least, will not make the situation worse. While the non-specialist may not understand the details of the TQ analysis, they can follow the steps and grasp the magnitude and assess the thoroughness of the evaluation.

1.2 Software Quality

In order to understand TQ, it is necessary to discuss the nature of software quality and software quality assurance to define the scope of concerns that must be addressed. The first question we have to answer is, "What is software quality?"

Quality may be viewed from multiple perspectives [4]. For example, software development is concerned with process quality. Software products (code and documents) exhibit product quality characteristics. Both process and product quality must address conformance to standards. During operation, quality attributes such as safety, security, and usability are important, as are other emergent properties. Customer value deals with customer satisfaction and perception of benefits.

This discussion focuses on software, but applies to software-dependent systems in general. A similar model [5] also includes the factor of service quality, which may be important for some products.

The software engineering process may be more or less capable of producing a good product. Process quality may be defined in terms of process standards or process reference models based on "best" (or at least "good") practices. Of course, in order to provide benefit, these practices must be properly implemented by the developers.

The software product itself may be evaluated in terms of conformance to appropriate product standards. The process perspective has proven to be more important for software quality than for typical manufacturing industries because software product standards have proven harder to agree upon, and logically, different standards are needed for different products. Software product standards typically define characteristics that are associated with successful operational use and maintenance of the software. Software product standards commonly focus on user interface and networking issues. Data may also be considered a software product that can be standardized for specific applications such as maps.

The operating software may be evaluated in terms of the quality of service provided, such as security, safety and usability. As emergent properties, these qualities are not amenable to standards. They must be evaluated in the overall systems context. For example, the safety of a software component depends on how the user interacts with it as well as the mechanical equipment attached to it. Conformance to product standards (such as "no single point of failure") cannot ensure safety, but can make it easier to achieve.

While safety is a special concern for offshore vessels, most safety analyses assume that software is 100% reliable, but reliability also depends on the context of use. Security has largely been ignored in this industry. For example, many suppliers deliver software with a hard-coded password. Nevertheless, quality (as freedom from defects and nonconformances to requirements), is a pre-requisite for both safety and security. Research shows that 50% of all security vulnerabilities are due to software defects [6]. Software quality has an immense effect on the customer's perception of the operational product, even if it is not the only factor.

The customer's satisfaction is based on their experience with the product as compared to their expectations of it. Satisfaction is a state of mind, not a physical state. Managing customer expectations is as important in achieving customer satisfaction as managing operational quality. Establishing standards for customer satisfaction is very difficult.

As mentioned above, the software process and software product are the views of software quality that are most amenable to standards. How do standards fit into software quality assurance? Standards provide expectations of activities to be performed or outcomes to be achieved. Satisfaction of these expectations can be confirmed by direct observation (verification) or quality assurance.

2. The Technology Qualification Process

Technology Qualification (TQ) is a process for demonstrating that a product containing new technology is safe, reliable, and fit for purpose. Figure 2 shows the an overview of the TQ process. TQ usually is performed by the innovator with the support and guidance of an independent party. In the maritime and offshore industry, this is typically a classification society [7]. Figure 2 is loosely based on the recommended practice for TQ by DNV GL [8].

Figure 2: Overview of the TQ process

The main steps of TQ are as follows:

- **Technology Assessment**: A study of the technology to identify the new and unproven parts requiring qualification. Establishment of product/mission objectives.

- **Technology Risk Assessment**: Identification of risks and potential mitigations associated with the new technology. Often a Failure Mode and Effects Analysis [9] is conducted in this step. Critical Performance Factors are defined. If data security is a concern, sensitive variables are identified [10].

- **Technology Qualification Planning**: Plan tests, analyses, and measurements to determine if the product is safe, reliable, and fit for purpose. A software/systems development process model often serves as a framework.

- **TQ Plan Execution**: Perform the tests and analyses in the plan. Collect the planned data.

- **Performance Evaluation**: Analyze the results of tests, analyses, and data collection. Calculate reliability and satisfaction of critical performance factors. The results are prepared as a report.

2.1 Software Technology Qualification

Whitfield [11] provides case studies of technology qualification for hardware. The innovation of the authors' work is extending the concept to include software. TQ of software-intensive products poses some special challenges. Exhaustive testing of software is problematic, so qualification must occur in parallel with development. One aspect of this is ensuring that the software is developed in accordance with accepted best software engineering practices.

2.1.1 Software Process Assurance

Many schemes for the evaluation of the robustness and dependability of software processes have been proposed in recent years. TQ yields an assessment of a product, not an estimate of the suppliers' likely performance, as often done.

Figure 3 summarizes key elements of the standard developed by DNV GL for assessing marine systems development [OS D-203, Integrated Software Dependent Systems (ISDS)][12]. DNV GL published the first version of this optional class notation in 2010; this standard was revised in 2011 and 2012, based on feedback from owners, shipyards, and suppliers.

ISDS defines a set of best practices for software engineering and systems integration, based on generic (not industry specific) international standards. Principles of ISDS are discussed in [13] and illustrated in Figure 3. Conformance to this standard is monitored via audits and reviews by the classification society.

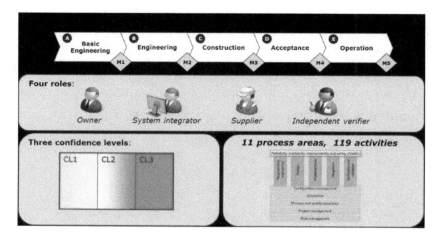

Figure 3: Key elements of ISDS (from [13])

There are more than 100 required activities, including peer reviews, milestone meetings, and requirements traceability data. Every activity is assigned to one or more roles in one or more lifecycle phases. The number of activities depends on the confidence level desired: the highest confidence (level 3) requires all activities, while the lowest confidence level requires only about 50 activities.

ISDS is similar in content to the CMMI [14] used in the defense industry. However, the application is substantially different. The CMMI is intended to be used in supplier selection prior to project start. However, the selected supplier does not always conform to the CMMI for the contracted project. In contrast, ISDS is used to monitor project execution. The classification society conducts audits and reviews during development to confirm that the engineering practices required by ISDS are executed at the appropriate time. Card [4] provides a summary of ISDS applications.

2.1.2 Software Novelty/Innovation

Most complex systems are a mix of established and new technology. Technology Qualification focuses on the new/innovative technology. Novelty of the software functions is evaluated using a four-level scale as shown in Table 1.

Novelty Level	Criteria	Example
1	Existing code previously used for this application	Ballast trim, Robotics framework
2	Existing code used in other application	Pattern recognition
3	New code, Existing algorithm	Communications protocol
4	New code, New algorithm	Pipeline following

Table 1: Software Novelty Levels for TQ

The amount of qualification/verification required increases with the Novelty Level of the function. The example column is based on an application of TQ to an autonomous underwater vehicle (AUV). For Remotely Operated Vehicles (ROVs), the hardware and some software components are largely proven. Novelty Level 1 software generally meets the requirements of ISDS for "proven in use" software, including:

- Systematic problem tracking and resolution

- Rigorous configuration control

- Multiple years of successful use with multiple deployments

Novelty Level 1 software does not require focused assurance separate from overall systems testing and evaluation.

2.1.3 Software Reliability

The TQ process involves an evaluation of the level of quality or reliability that has been achieved in the product. Assessing software reliability remains a challenge for practitioners and researchers. The authors have had success with an analysis approach based on the Weibull family of models [15]. The Weibull model is suited for processes where defects are continually being introduced and removed from the system, but at different rates. Figure 4 shows an example of a Weibull analysis of data from testing and commissioning of a control systems for marine navigation, following a simplified approach to Data Envelopment Analysis [16]. The Weibull curve represents the theoretical optimum performance of this verification process.

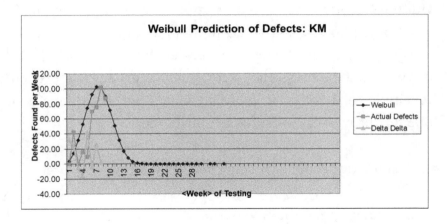

Figure 4: Example of software reliability modeling

However, in any given testing interval there are likely to be inefficiencies, e.g., holidays, delays, poor planning – so the actual performance tends to fall below the optimal performance. In this case, regression on the actual data (e.g., [17]) will not give us the correct (frontier) Weibull model. Instead, we impute the Weibull parameters using the following criteria:

- The observed peak of the distribution should coincide with the peak of the theoretical model.

- The slope of the observed data should match the slope of the theoretical model. The Delta Delta line in Figure 4 shows the differences in the slopes of the two curves.

- Both curves pass through the origin.

Thus, we can identify Weibull parameters to estimate the tail of the frontier model. The result shown in Figure 4 indicates that many defects remain at the end of this phase of testing. Additional testing is needed to reduce the defect level and increase reliability. Figure 4 shows that 404 defects actually were found through 10 weeks of testing. The Weibull curve indicates that 749 defects should be found through 16 weeks of testing. Subsequent testing discovered additional defects consistent with the gap between the curves shown in Figure 4.

3. Examples of Software TQ Application

This section discuss two examples of projects where TQ was performed. The TQ activities were performed by the product developer under the guidance of DNV GL. Unfortunately, intellectual property considerations prevent discussion of full details of these projects.

3.1 Autonomous Underwater Vehicle

The mission of the autonomous underwater vehicle (AUV) project was to perform inspections of underwater structures such as pipelines and jackets. The hardware technology is well-established as remotely operated underwater vehicles (ROV) are in common use. It is the software that makes the systems autonomous and requires qualification. A major concern of the potential adopter is that the AUV might collide with and damage valuable underwater structures.

Figure 5: Aquabotix Hydroview ROV

Figure 5 shows a popular ROV, the Aquabotix Hydroview [18]. The ROV is connected to the surface by an umbilical that provides reliable power and communications, but that also limits the ROV's mobility.

Critical performance factors for TQ of the AUV (Step 3 of Figure 2) included:

- Battery life

- Memory capacity for images

- Accuracy of trajectory following

If this performance can achieved with demonstration of safety and reliability, then the AUV can be accepted as a viable alternative to the ROV, but without the limitations of the umbilical.

The Weibull approach of Figure 4 was used to evaluate sufficiency of testing. As a result of the TQ effort, the customer agreed to allow sea trials of the AUV within their facility to collect additional data.

3.2 Remote Acoustic Controller

Another TQ example is an acoustic controller for blow-out preventer (BOP) systems. The BOP is a system of valves that is intended to shut down an oil well in case of an emergency. The acoustic BOP controller allows the operator to control the BOP even after evacuation of a distressed drilling rig.

A major concern of the potential adopter of the acoustic BOP technology is that emergency procedures employing the acoustic controller could prove inadequate if the controller failed to operate reliably. The TQ effort addressed this broad concern. Unlike the AUV example, the scope of this TQ effort included software as well some acoustical hardware components. The result of the TQ effort facilitated the adoption of this device as part of offshore oil well safety systems.

3.3. Summary

The cost and benefit of the new technology was easy to estimate in both of these examples. However, there is substantial risk of loss in case of the failure of the technology, motivating the investment in technology qualification. TQ did not guarantee that these products would work as intended, but it helped the adopter to understand and mitigate the risks involved with adoption.

The extension of the Internet of Things (IoT) involves integrating software and computers into more everyday products. As this continues, products posing health and safety risks will be included. Processes such as TQ will be needed to ensure that these components are safe, reliable, and fit for use in the IoT environment.

References

1.   E. Rogers, Technology, Adoption Lifecycle model: Diffusion of Innovations, Free Press, 1962.

2.   B Hall and B. Khan, Adoption of New Technology, New Economy Handbook, 2002.

3.   S. Khairat, D. Marc, W. Crosby, and A. Al Sanousi, Reasons for Physicians not Adopting Clinial Decision Support Systems, Journal of Medical Informatics, April 2018.

4.   D. Card and L.G. Chua, Ensuring Software Reliability, Safety, and Security, Proceedings: RINA international Conference on Computers in Shipbuilding, 2017.

5.   A.R. de Rocha and G.H. Travassos, QPS – Modelo de Referencia para Avaliacao de Produtos de Software, Universidade Federal do Rio de Janeiro, 2018.

6.   C. Woody, Predicting Software Assurance Using Quality and Reliability Measures, Software Engineering Institute, 2014

7.   Classification Society, Wikipedia, 4 January 2018.

8.   Det Norske Veritas (DNV), RP A-203, Technology Qualification.

9.   D.H. Stamatis, Failure Mode and Effects Analysis, ASQ Press, 2003.

10.   P. Ogale, M. Shin and S. Abeysinghe, Identifying Security Spots for Data Integrity, Proceedings: IEE Computer Software and Applications Conference, 2018.

11.   S. Whitfield, ADNOCS Case Studies Illustrate Importance of Technology Qualification Process, Journal of Petroleum Technology, August 2017.

12.   Det Norske Veritas (DNV), OS D-203, Integrated Software Dependent Systems (ISDS), December 2012.

13.   Card, D., A Software Integration and Process Model for Offshore Vessels, Proceedings: Asia-Pacific Software Engineering Conference, 2013.

14.   Chrissis, M., M. Paulk, and S. Shrum, Capability Maturity Model - Integrated, Addison Wesley, 2003.

15.   Card, D., Managing Software Quality with Defects, Proceedings, IEEE International Conference on Software and Applications, 2002.

16.   Beasly, J.E., Data Envelopment Analysis, OR Notes, Brunel University, http://people.brunel.ac.uk/~mastjjb/jeb/or/dea.html, 2018.

17.   D.R. Dola, M.D. Jaybhaye, and S.D. Deshmukh, Estimation of systems Reliability using Weibull Distribution, International Proceedings of Economic Development and Research, 2014.

18.   Aquabotix Hydroview, www.westmarine.com, 2018.

About the Authors

David N. Card is an independent consultant working in the areas of software quality assurance for aerospace, automotive, and marine systems. He is also a research associate of the Experimental Software Engineering Group of the Universidade Federal do Rio de Janeiro, Brazil. Previous employers include Computer Sciences Corporation, Det Norske Veritas, and Lockheed Martin. Mr. Card is the author of many articles and two books on the topics of software quality, reliability, estimation, and performance management. Contact him at card@computer.org.

***

Michelle E. Novaes-Card is a Reliability Engineer with Northrop Grumman Corporation (NGC), in Melbourne, Florida. She received an Innovation Award from NGC in 2016 for her work modeling aircraft availability. Michelle attends the Florida Institute of Technology where she is pursuing a Master's degree in Operations Research. Previously, she received a B.S. in Mathematics from the University of Central Florida. Contact her at Michelle.novaes-card@ngc.com.

#

The mighty Saturn V lifts off for the Apollo 11 mission with astronauts Neil Armstrong, Michael Collins, and Buzz Aldrin on July 16, 1969, from KSC Launch Complex 39A. (NASA)

A Systems Engineer from Left Field

James Collins

Abstract

Where do systems engineers come from? How do they arise and how do they develop? Is there one path or are there many? This chapter explores the experience of the author and argues that there are many paths, but the common denominators are the development of a systems view, and a breadth of experience with multiple technical disciplines and projects.

1. The 1970s

Not all systems engineers began their careers as engineers. Some of us migrated into systems engineering from other STEM disciplines, like the physical sciences or mathematics. There are even a few psychologists by training who moved into SE as human factors specialists.

In my case, I started out in physics and astronomy (B.S., M.A., Ph.D.). I planned to go into academic research and teaching, but the prevailing winds didn't blow my way. After a couple of years in academia, I returned in 1976 to the United States government employer for whom I had worked in the summers as an undergraduate in my home state of Indiana.

Since I acquired skills in computer programming, modelling, and simulation in graduate school, that made me a candidate for working on simulation and embedded systems at my government employer. Of course, this was when most aerospace-related programming meant FORTRAN or assembly language. An early assignment involved writing tools to support an embedded computer system, but the tools had to be written in a then-modern language called BLISS that looked nothing like the FORTRAN I had previously used. This was my first exposure to a more modern computer language designed for implementing embedded systems software and was a valuable break from my academic programming experience. This assignment also provided my first exposure to microcomputer assembly language programming and hardware/software interfacing.

Another early assignment was performing missile launch envelope simulations. I was part of a small team that moved the simulation from vacuum tube analog computers and slow mainframe digital simulations to minicomputer simulations written in FORTRAN. Although the older analog simulation was well established and operated in near real time, and the mainframe simulation offered very accurate results, my team argued that the digital simulation on a minicomputer would produce better data than the analog computer and faster turnaround than the mainframe – and it did. I was then assigned to develop and run the hybrid (analog/minicomputer) lab that hosted the that simulation, but that lab rapidly evolved into a minicomputer-only lab.

Although management expressed a desire to maintain and update the analog computers, we eventually argued successfully for their retirement. We noted that the newer engineers joining our organization usually came with experience in simulation on digital computers rather than analog computers, and that the rapidly improving speed and capacity of digital computers made the historical advantages of analog computers much less relevant. The minicomputers were also less expensive to buy and maintain than the analog or hybrid computers that they replaced.

3. The Early 1980s

Around 1980, our site technical management became aware of the rise of digital avionics development and integration facilities at other government and aerospace contractor sites. I was tasked with building one for local use as a competitive tool and with building a team to implement it and support other projects in using it. Over the next several years, I gathered the team and we specified and integrated a simulation lab based on large-scale minicomputers and real time graphics.

Throughout this period, when management would discuss assignments and job performance, there was always the question of what category of "engineer" I was. Occasionally, someone would suggest a category like "systems engineer," but that wasn't one of their standard choices in the 80's.

4. The Mid 1980s

With the laboratory up and running, in 1985 I moved from my federal government employer in Indiana to a private aerospace company in Florida to support a prospective new systems integration contract with a required digital avionics integration lab. It was then that I joined a systems engineering department for the first time. I was also categorized as a "systems engineer," but still without a systems engineering degree or any engineering degree at all. However, my new employer recognized the need for people to perform the types of activities associated with systems engineering, such as analysis and simulation, requirements definition, specification writing, proposal support, design reviews, customer technical liaison, and interdisciplinary coordination.

The contract and its laboratory development did not materialize, but I did work on three other projects that taught me a lot about requirements, specifications, interface control documents, technical coordination between different organizations, and customer technical liaison. The first of these projects required me to compile an interface control document (ICD) for an integrated avionic systems after the fact. My efforts started with a large stack of interface diagrams and notes from the development team that needed to be collated and reconciled into a deliverable ICD.

A second project required me to develop requirements and an architecture for a programmable digital test station to support the production test and maintenance of that same avionic systems and then to oversee the station's development. The third project required me to work as a technical liaison with another contractor using technology that my employer had developed as part of a major competition. I learned a lot about practical systems engineering from all three projects, and worked with other good engineers practicing (and learning) systems engineering on the job.

5. Late 1980s, the 1990s, and the Early 2000s

In 1987, I moved to another Florida company with an eye toward applying my experience in digital avionics integration laboratories. I did have role in developing and operating such laboratories for the first years of work for my new employer, but a funny thing happened along the way.

After a few years of working on the laboratories for a major systems program and interfacing with other departments frequently, including our security department, I was offered the role of day-to-day technical liaison and providing cooperation between engineering and security. This role bridged the security organization, whose expertise was in the letter of government security requirements, with the technical demands of the company's engineering organizations. It was part of a program to foster a cooperative culture of security awareness and compliance. In this role, I engaged with the company's security team and with the federal government security representatives, and learned their perspectives and requirements.

My management in engineering once told me that my job description was to "keep security problems off management's desk," and this goal was my focus for over 15 years. In the later years, my role expanded to include engineering support to export technical compliance using similar methods, and I often joined proposal teams that included foreign suppliers, subcontractors, or partners. Both latter assignments required the ability to listen and communicate across a range of technical and non-technical disciplines, and to resolve uncertainties and disagreements between parties with differing priorities while protecting my team and my employer from information control violations.

6. The Mid 2000s Until Today

Starting in 2003 and well past mid-career, I participated in a Master of Science in Systems Engineering (MSSE) program sponsored by my employer and offered by the Florida Institute of Technology (FIT). My goal was to gain a deeper and better organized understanding of systems engineering than I had accumulated over the years working on many projects. After three years of study, I finally became a systems engineer by education and degree. Since then, I have taught classes in the MSSE program at FIT.

I retired from my final aerospace employer in 2012 and continue to teach systems engineering, but no longer as a systems engineer from left field. I warn my students about my unusual background and to expect some examples and analogies from astronomy and space science, as well as other engineering disciplines.

7. Career Reflections

I really did set out to be an academic, and the required redirection of my career into engineering was a disappointment for a long time. I often felt like an outsider working in engineering organizations among people who were actually educated in engineering specialties. Mind you, I did attend a college (Rose Polytechnic Institute) from which it was difficult to graduate without at least some engineering education, and many of my professors were very practical in their outlook and provided valuable guidance and direction. Later, in graduate school, the guidance was more specialized and focused on astronomy and astrophysics as it was supposed to be, but my professors did provide opportunities to expand my experience in research and computational skills that served me well later along my new systems engineering path.

I also began teaching college classes and became comfortable with organizing material for presentation and speaking before groups of people. These latter two items are often overlooked skills needed by systems engineers. I had the benefit of mentoring by several senior engineers when I worked for the U.S. government, and they mentored me in organizing and scheduling work and people and provided a range of technical assignments that gave me insights that I had not acquired in my academic years into working with teams.

When I worked for the private aerospace companies, I was given more responsibilities, and I learned some key lessons about bringing a large effort to a successful conclusion with the mentoring of several senior colleagues. I also came to appreciate the technical needs of other organizations and specialties and acquired knowledge of their unique problems and problem-solving methods. I owe a great deal to many more senior colleagues who provided guidance and opportunities to grow systems engineering skills and who knew when I needed a nudge to take the next step along the path.

When I applied for the MSSE program, I still recall speaking with the FIT department chair in charge of the program and vaguely apologizing for not having the nominal required engineering undergraduate degree. He assured me that the breadth of my work experience on top of my undergraduate and graduate education should prove quite adequate, and he was right.

***

Today, I understand that most aerospace employers appreciate the value of formal education in systems engineering, and encourage or insist that their systems engineers have such preparation. Systems engineering has matured into a well-recognized discipline with its own skill set and body of knowledge that students and practicing engineers can study and apply. Many government and private contracting organizations also expect their contractors' staff to include experienced systems engineers in order to improve the chances for meeting cost, schedule and performance.

Could a systems engineer still come in from left field, as I did years ago? Yes, I believe so, but formal education in the methods and perspectives of systems engineering before mid-career is very valuable and should accelerate the progress of any STEM graduate from their original specialty into the broader world of systems engineering.

Still, systems engineering isn't usually taught at the undergraduate level because of the need for experience in a STEM specialty, and a broader understanding and appreciation of a range of other specialties and of systems management as well. I believe that much of this cannot be taught in a classroom; it must be learned through experience. Of course, I would recommend that a prospective systems engineer not start out as an astrophysicist – that really is from left field!

About the Author

James Collins retired from a major aerospace contractor in 2012 after 25 years and has been teaching systems engineering classes at Florida Institute of Technology in Melbourne, Florida since 2006. He holds a BS in Physics (1969) from Rose Polytechnic Institute (now Rose-Hulman Institute of Technology) in Terre Haute, Indiana, an MA in Astronomy

(1972) and a PhD in Astrophysics (1975) from Indiana University in Bloomington, Indiana, and a MS in Systems Engineering from Florida Institute of Technology (2005). He has published multiple papers in professional journals and holds a United States patent for a security interface device. Contact him at jcollins@fit.edu.

# # #

FLORIDA TECH

FLORIDA'S **STEM** UNIVERSITY®

The Florida Institute of Technology opened in 1958 as Brevard Engineering College by Dr. Jerome P. Keuper, a physicist who worked at Cape Canaveral during the start of the space program. The university's name was changed to "Florida Institute of Technology" in 1966. Three Florida Tech faculty members have been to space (Buzz Aldrin, Samuel Durrance, and Winston Scott) and five of Florida Tech's alumni have served as Space Shuttle astronauts. The university offers graduate degrees in systems engineering and engineering management. (Florida Tech)

Systems Engineering in Construction

Troy Nguyen, Jack Dixon,
Aldo Fabregas, and Peter Zappala

···

Abstract

This chapter focuses on the application of systems engineering (SE) methodologies to complex large-scale construction projects, where logistics, coordination of different teams, and integration of various systems present many challenges. A case study is included that illustrates the SE processes employed in the design and development of an intelligent, scalable, zero-energy commercial office building to achieve cost and schedule objectives.

1. Introduction

The demand for residential and commercial construction, and opportunities involving large-scale infrastructure development, is expected to continue to grow steadily. Today's construction projects are complex, have stringent cost constraints, and demand long-term sustainability. The scale of the projects, the number and types of requirements, the multiple technologies integrated into buildings, and the many demands of stakeholders create great challenges. Modern facilities require integration of new technology and advanced computer systems with the more traditional elements. Given these challenges, the construction industry can benefit greatly by adopting the systems engineering (SE) approach to the development.

In 2004, the construction industry comprised 4 percent of the U.S. GDP [1]. Residential and commercial development in the U.S. is expected to grow steadily over the next 20 years. With 614,387 bridges, and almost four in 10 of which are 50 years or older, it is expected that major infrastructure upgrades will be planned and/or underway [2]. Current environmental policy promotes energy-efficient construction materials and methods, in addition to requirements for more efficient land development techniques and mixed construction processes. Advances in materials and technology, such as the Internet of Things (IoT), summed with increased user expectations for usability and services, requires additional stakeholders and subsystems to be considered in construction projects.

Systems engineering (SE) is a set of proven concepts that can support large-scale and complex construction projects where logistics, reliability, coordination of different teams, and integration of various systems are often challenging issues. Understanding these basic principles is an essential part of a construction manager's toolbox. In fact, based on negative experience with recent large scale construction projects, many government and industry entities require systems engineering capability as a prerequisite to construction contract bidding. This chapter provides a convergent high-level viewpoint between SE and construction management. A generalized lifecycle of construction project is mapped to SE processes to be treated with traditional (e.g., interface analysis) and modern SE techniques (e.g., model-based systems engineering). The application of SE is demonstrated through the design of a zero-energy commercial office building.

2. Systems Engineering Overview

The International Council on Systems Engineering (INCOSE) defines systems engineering as an "interdisciplinary approach and means to enable the realization of successful systems. It focuses on defining customer needs and required functionality early in the development cycle, documenting requirements, and then proceeding with design synthesis and systems validation... Systems engineering integrates all the disciplines and specialty groups into a team effort... Systems engineering considers both the business and the technical needs of all customers with the goal of providing a quality product that meets the user needs." [3].

An extensive body of experience collected from industry suggests that nearly two-thirds of software development projects in the U.S. fail, either through cancellation, overrunning their budgets, or delivery of software that is never put into production [4]. The reasons for these failures include a lack of stakeholder involvement, no clear statement of requirements, no project ownership, no clear vision and objectives, and lack of planning.

2.1 The Systems Engineering Solution

While the study [4] was based on software development, many of the failure causes are applicable to construction projects. For example, work hand-off from designers to contractors and among contractors is always subject of errors due to miscommunication of requirements or lack of verification. SE can make improvements in this area by establishing good practices for designers to produce consistent sets of requirements, functional arrangements, and design solutions. The end goal is to improve productivity and at the same time provide deliverables that satisfy the product's end purpose.

Since today's construction projects can be large and complex, and can include multiple advanced technologies, the systems engineering approach contains the methodologies to bring these projects to successful completion. The emphasis on satisfying divergent needs of customers and stakeholders requires a focus early in the design process on defining the requirements in detail, documenting them, and establishing the necessary functionality to satisfy these requirements.

Using the SE approach guarantees that the complete picture is considered during design synthesis. All aspects of the project throughout its expected life, including Operations, Performance, Test, Manufacturing, Cost & Schedule, Training & Support, and Disposal must be considered in the design phase. Therefore, numerous specialties must participate in and be completely integrated into the design process. The SE process is the right approach to ensuring a successful construction project.

2.2 Systems of Systems

In recent years, systems engineering has evolved in numerous ways, but in many complex cases it has evolved into a discipline of systems of systems (SoS). There are various definitions of SoS, but Mo Jamshidi's definition is one of the more straightforward ones: "Systems of systems are large-scale integrated systems that are heterogeneous and independently operable on their own, but are networked together for a common good" [5]. Examples of SoS are space exploration systems, many military systems, communication systems, and air transportation.

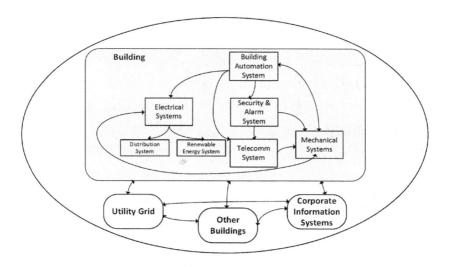

Figure 1: Buildings as systems of systems

As shown in Figure 1, construction projects can be considered systems of systems. As an example, large commercial buildings, each designed to serve different purposes (i.e., offices, retail space, or mixed use) are interconnected, and their operation and maintenance are coordinated under a corporate umbrella for economic purposes. Advances in networks and information technology service have also enabled these networks of buildings to interconnect with other entities, such as electrical utilities for energy management and/or use of electricity.

2.3 Scalability

Scalability is the ability of a systems to grow in order to handle increased demand. The systems of interest may be a network, an organization, a business, a facility, or a large construction project. A scalable building or a highway is adaptable to the growing demands of the consumer.

Scalability requires that the construction project be designed from the beginning with scaling in mind. Through the SE process, the project architects, engineers, and stakeholders should evaluate all possible expansion scenarios during the initial design efforts. Specific requirements for scalability must be defined for all the important dimensions. Every aspect of the systems must be carefully analyzed to determine where growth may be expected, and solutions must be incorporated into the design to accommodate that expected growth.

3. The Systems Engineering Process

The systems engineering process comprises a combination of management processes and technical processes that are executed throughout the project's lifecycle. The management processes include planning, configuration management and control, management of interfaces, management of technical data, requirements management, risk assessment and management, and decision analysis. The technical processes include development of stakeholder requirements, analysis of requirements, design, installation, validation and verification, operation, and disposal.

The SE process model has been revised several times over the years as systems have evolved. The following sections highlight some of this evolution and how the systems engineering process can be applied in the construction industry.

3.1 Traditional Systems Lifecycle Processes

A system's lifecycle is described by the different phases that the system goes through from inception to disposal. This is often referred to in terms of the human lifecycle, such as planning for a systems from "birth-to-death" or "cradle-to-grave." The typical phases of systems development are: Concept, Development, Installation & Acceptance, Operations, Planned Product Improvement, and Decommissioning & Disposal. Figure 2 depicts these six phases.

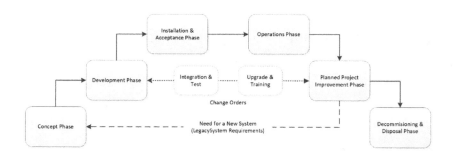

Figure 2: Systems lifecycle phases

During the **Concept Phase**, a need is recognized and a concept is developed for a systems that will meet this need. At this stage, a concept of operations (CONOPS) is proposed to satisfy those needs. Often the CONOPS documentation includes early elements of the systems architecture, which are then expanded in the lifecycle's next phase.

During the **Development Phase**, a requirements analysis is completed and the systems architecture is defined. Following the development phase, a procurement process is used to select the contractor who will build and deliver a systems that meets the defined requirements.

During the **Installation and Acceptance Phase**, the system is tested against the requirements and will often have a third party perform an independent test of the system, referred to as independent validation and verification. After the systems is installed, the owner will perform an approved system acceptance test. Once the system successfully passes all required tests, the contractor transfers ownership of the system to the owner.

During the **Operations Phase**, the system is operated in its specified environment. Any problems that arise are usually fixed by the contractor providing system support, or by the owner's personnel the contractor has trained to maintain the system.

If upgrades are warranted, the **Planned Product Improvement Phase** will be initiated. The procurement process should plan for an eventual system upgrade to incorporate new technologies.

Eventually, the systems will become worn out or obsolete, at which time the **Decommissioning and Disposal Phase** begins. When the system no longer meets the stakeholder needs, or has become too costly to maintain, a cost/benefit analysis will indicate that it is time to replace the systems and the entire systems lifecycle process is repeated.

Decommissioning involves switching the operations from the legacy system to the new system in a way that allows the owner to revert back to the legacy system if the new system doesn't function reliably. The disposal process must be addressed in the early stages of the system's original concept phase, because it could be an expensive and time-consuming process if not adequately planned for.

3.2 The Lifecycle of Construction Projects

The lifecycle of construction projects is similar to other product lifecycles, but differs in significant ways. Typical construction phases include: Concept, Promotional, Design, and Construction. However, when thinking of construction projects from a systems perspective, for completeness, one should add the Operation & Maintenance (O&M) phase and the Disposal phase, which are typically the responsibility of the owner. Figure 3 is a representation of the construction lifecycle.

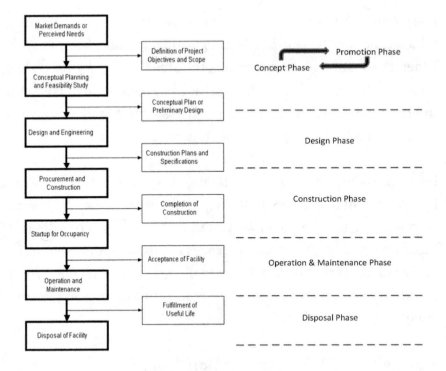

Figure 3: Lifecycle of construction projects

During the **Concept Phase**, the owner/investor decides that a building is needed for some particular purpose. During this phase the following activities are undertaken:

- Determine the construction location

- Acquire property through a developer or realtor

- Perform a land survey

- Determine property boundaries

- Work with city/county planner to determine permitted land use

- Draw conceptual sketches (by architect)

During the **Promotional Phase**, presentations will be made to interested parties, which may include bankers, investors, homeowner association, and the county/city development/planning board. The goal of this phase is to maximize the benefits to the owner (e.g., more parking spaces, less stringent requirements on exterior features that cost extra, energy efficiency, and so on).

During the **Design Phase**, the architect develops the architectural details, including building exterior/interior features, and site landscaping. The architect coordinates other disciplines, such as Civil, Structural, Electrical, Mechanical, and Fire Protection to design the building systems. Outputs of the **Design Phase** are a set of construction documents (CDs) that include drawings, specification documents, and schedule requirements. The CDs are then issued in a request for proposal to obtain bids for the job.

During the **Construction Phase**, the selection of a general contractor is made. The general contractor's responsibilities include obtaining building permits from city/county, providing the labor and materials necessary to meet contract terms, scheduling, supervising and managing subcontractors and material deliveries, and scheduling inspections with city/county building departments. The general contractor also works with the architect and engineers to clarify or revise working plans as necessary to make sure the construction meets design intent.

During the **Operation & Maintenance Phase**, the owner manages the finances of the building operation (e.g., lease revenue, operating costs). The owner also maintains the building and building systems to ensure the comfort of occupants and the efficient use of resources.

During the **Disposal Phase**, the owner must decide whether to retire the building due to aging or due to the need to replace it with something newer. The owner will most likely need to conduct an environmental impact assessment for any substantial size project. After the environmental impact assessment is complete and accepted by the proper authorities, the owner will demolish the structure while controlling the pollution and transport the debris for proper disposal in accordance with environmental regulations.

3.3 Traditional Systems Engineering Process Models

There have been numerous systems engineering process models developed over the years. The three most-used ones are briefly discussed in this section.

3.3.1 Waterfall Models

In the waterfall model, lifecycle development begins the gathering of requirements and domain knowledge and ends with systems deployment, maintenance, and eventually, retirement. Each phase of development is sequentially completed, via formal review, before the next phase begins. A variation is the Incremental Waterfall Model, where development of a systems is a series of versions or increments. At each increment, a subset of functionality is selected, designed, developed, and implemented.

3.3.2 The Spiral Model

The spiral model corresponds to a sequence of waterfall models and is a risk-oriented iterative enhancement. It recognizes that implementation options are not always clear at the beginning of a project. An implementation option may be uncertain, for example, because it is critically dependent on a technology still under development.

3.3.3 The "Vee" Model

The "Vee" model depicts a top-down development and bottom-up implementation approach. This model is most closely related to the different phases of typical construction projects. On the left side of Figure 4, decomposition and definition descends as in a traditional waterfall model. On the right side of Figure 4, integration and verification ascends as successfully higher levels of units, assemblies, and subsystems are integrated and verified, culminating at the systems level.

The "Vee" model is a composition of three layers, or perspectives, of the system in increasing engineering detail:

1. User's Perspective: This is the view of the user who is interested in presenting a list of requirements and receiving a finished product that meets the requirements.

2. System Engineer's Perspective: This perspective encompasses the architectural details which addresses the decomposition of the system-level specification into system design and subsystem specification and designs, paired together with built and tested subsystems, and finally the tested system.

3. Contractor's Perspective: This perspective covers the implementation process that is normally performed by contractors and/or subcontractors. In practice, the contractor's perspective is associated with component specifications and designs with fully tested components

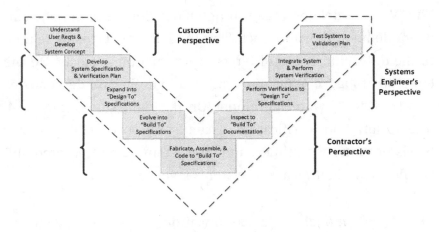

Figure 4: The "Vee" process model

The "Vee" model leverages the advantage of waterfall model in illustrating the evolution of user requirements into preliminary and detailed designs in the top-down manner. But it also accommodates integration and verification of systems components through subsystem and systems testing using a bottom-up path that tends to allow design reuse.

3.4 A Systems Engineering Process Model for Construction Projects

In recent years, the construction industry has begun to adopt systems engineering principles that have been the foundation of success in the defense and aerospace industries. The "Vee" model represents the best starting point for construction projects.

The "Vee" model represents the sequence of steps in a construction project development and includes the activities undertaken and the designs created during development. The model shows the specification, the design, the build, and the verification & validation. In addition, it details the decomposition of the systems to the subsystems and the components to the elements. It illustrates the verification and validation against the requirements.

To be able to properly manage the complexity of a construction project, the top-down approach is essential. Depending on the complexity of a system, this process may be repeated multiple times as the design evolves from the top level to the lowest level. The iterative engineering process is repeated at each lower level until the level of detail is sufficient for the design to be constructed. Figure 5 shows the "Vee" model adapted to the construction industry.

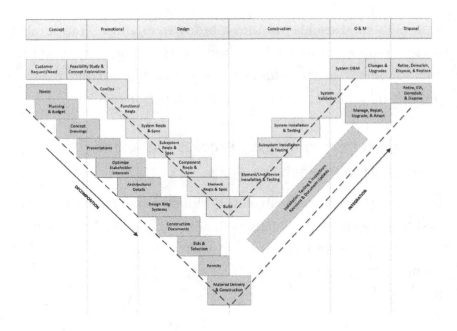

Figure 5: A "Vee" model for the construction industry

3.5 Applying Model-Based Systems Engineering

Model-based systems engineering (MBSE) supports the analysis, specification, design, and verification of activities by creating a digital representation of the concepts involved in the system design. The system model starts in the problem domain from the stakeholder's perspective (e.g., desire of an energy efficient building) and evolves through the solution domain (e.g., electrical systems architecture).

A SE process uses a methodology that determines project management aspects and quality gates for design artifacts at each stage (e.g., validation plan). To apply MBSE, a method, a language, and a tool are required. MBSE methodologies include Object Oriented Systems Engineering Methods (OOSEM) [6],

IBM Harmony [7], SysMod [8] and Magic Grid [9]. The Systems Modeling Language (SysML) is one of the most common languages used to represent systems design.

4. Construction Project Management

A construction project is a temporary and unique set of operations needed to accomplish one goal, such as building a commercial building, a bridge, or a power plant. A project team is led by a project manager and composed of individuals from many technical and non-engineering disciplines, and from different geographical areas. All these people and their functions must be properly managed in an effort to deliver an acceptable product, on time, and within resource budgets, with acceptable risk. Construction project management, therefore, is the application of knowledge, skills, and tools to meet the construction project's requirements.

4.1 Risk Management and Cost Control

Risk analysis is an integral part of the SE approach and is essential in all construction projects. Risk analysis is performed at the beginning of a construction project and continuously updated throughout the project. Doing this helps to foresee the risks in the future and to effectively deal with them, thus reducing the chances of failure or of large, unexpected cost overruns. Project risk management processes include identifying and analyzing project risks, and developing options to control and minimize threats to project objectives.

4.2 Configuration Management

Changes in construction projects always bring risks to costs or schedules. All changes to the design, even when small and insignificant, need to be evaluated for additional risk they may pose to the project and possible negative effects on the design. Changes become more risky if made late in the development process. It is important to specify complete, detailed requirements early in the development process to minimize changes.

However, regardless of how well requirements are developed early in the development process, change is inevitable. Therefore a vigorous configuration management (CM) process must be in place to ensure any changes that are made follow a strict review process, are made in a systematic fashion, and any additional risk is minimized.

In recent years, the construction industry has embraced Building Information Modeling (BIM) methodologies in the design, analysis, and implementation of construction projects. BIM encompasses the technologies and the processes that enable 3-D visualization of construction projects for design and analysis, and the ability to manage design changes resulting from collaboration among project stakeholders. Typical BIM software applications are capable of accommodating design changes and retaining the history of those changes in a permanent database, thereby providing an effective tool to manage and document the project changes and configurations over its lifetime.

4.3 Integrated Product Teams

Integrated Product Teams (IPTs) comprise stakeholders who plan, execute, and implement lifecycle decisions for the systems being acquired. When the team is dealing with an area that requires a specific expertise, the role of the member with that expertise will predominate. However, other team members' input should be integrated into the overall lifecycle design of the product.

In many organizations, the lead systems engineer on the project is the Integrated Product Team Lead (IPTL). Design engineers and stakeholders report to the IPTL, as do the other specialty engineers, such as reliability engineers, maintainability engineers, and supportability engineers. All members of the IPT will thus have an opportunity to interface with all the other members of the IPT on a regular basis.

By virtue of these regular meetings, the interfaces between multiple specialty engineering disciplines and non-engineering professionals can be addressed, and the exchange will help achieve a successful project.

5. Specialty Engineering

The design, development, and implementation of a complex construction project often requires the capabilities and expertise of many engineering specialties. The project manager is often assisted by engineers who are experts in the various engineering specialties, such as reliability, human factors, quality management, environmental engineering, and so on.

Integrating engineering specialties into a project team increases the expertise available to define the design requirements characteristic of these technical fields. Systems engineering ensures that the various engineering specialties perform their tasks efficiently, and that they are integrated into a project from concept design through systems installation and support.

Timely and accurate interfacing of all team members and project specialties is essential for a project's success. Effective and efficient interfacing is accomplished by maintaining and implementing detailed plans and schedules that integrate all project activities.

6. Plans and Planning

Successful project management and systems engineering management requires planning. Good planning requires collaboration between project management, stakeholders, and technical specialty experts to ensure that all interests are adequately addressed, all tasks and activities are defined, resources are sufficient, risks are identified, schedules are achievable, and timing of various events is appropriate.

6.1 Systems Engineering Management Plan

Implementing a SE process for the deployment of construction projects will increase the likelihood of a project's successful deployment. A key ingredient is the implementation of a systems engineering management plan (SEMP). A SEMP will help to improve the success rate of construction projects.

The SEMP is a plan that describes how the project will be technically managed. It alerts all project participants to the rules of the road. The SEMP is the top-level technical management document, and it should be created early in the concept phase of the project. It becomes a living document and should be updated throughout the lifecycle of the project. The development of a SEMP requires contributions from the project manager and the entire project team, as well as from technical experts from all specialty areas that can significantly impact the outcome of the project.

6.2 Other Plans

The SE process spans the entire systems lifecycle, from the beginning of requirements definition, through design, construction, support, planning for replacement, and eventually disposal at the end of the system's useful life. It is an iterative process involving project and technical management, procurement, supply, design, and evaluation at each level of the system. SE involves planning for all the various aspects of the project. This effort results in a set of plans for managing and controlling these processes.

Many types of plans may be created in conjunction with the SEMP, and many of these plans are directly applicable to a majority of engineering or construction projects. These may include general project management and other plans, such as: project management, systems integration and test, configuration and data management, quality management, subcontract management, cybersecurity, and so on.

7. Implementation of Systems Engineering

The key to a successful implementation of the SE approach is to identify systems requirements at the beginning of the project, track the requirements through the design process, and test the final product to verify that the requirements are satisfied.

7.1 Development and Analysis of Requirements

The definition and creation of requirements is a critically important function in systems engineering. If the requirements are wrong at the beginning of the project, it will have a ripple effect throughout the project's development.

Functional Requirements: During the earliest phase of a project, most of the top-level system requirements will describe system functions. The project's CONOPS will state the problem to be solved by the project and will create a description of the system as a collection of functions that are necessary to meet the stakeholders' needs.

Performance Requirements: After the top-level functional requirements have been identified, they are translated into performance requirements. Reviews with the stakeholders can be used to refine requirements and ensure that the interpretation and understanding of the requirements are consistent with the needs.

Requirements Analysis: Requirements analysis is performed throughout development of the system. The analysis begins with the preparation of system block diagrams. Functional block diagrams are created for the system. Analysis of the functions is critical to deciding if a function is to be allocated as hardware or software or a combination of both. Once the function has been allocated, the detailed performance requirements are developed and recorded in the appropriate top-level, subsystem, or component specification.

Requirements Database: It is imperative that throughout the development of the system the requirements are tracked. They must be linked back to the user needs. All project requirements should be managed using a database that illustrates the hierarchal structure of the system, links the requirement to the user need, provides change control information, and shows how the requirement is being met.

7.2 Developing Requirements for Construction Projects

Today's construction projects are complex and complicated. They are larger, have a greater number of requirements, must meet more stringent codes and standards, have more stakeholders, and have more technology integrated into the projects. Historically, the construction project was typically controlled by "The Architect," whereas modern projects require large teams of specialized personnel all working together to create the ideal solution.

Traditionally, construction projects were driven by regulatory terms found in codes and standards. Cost and schedule often also constrained the project. Relatively little emphasis was placed on the anticipated performance of the end product. Today, regulatory bodies, as well as customers, are placing more emphasis on performance of the end product. Therefore, it is becoming more important that performance requirements are developed upfront, managed throughout the project, and measured to ensure that they are met.

Godfried Augenbroe provides an adaptation of the systems engineering approach for construction [10]. He has proposed a "...design process for construction projects in which the design problem is reduced into smaller manageable functional units. For each of these units, a designer looks for a technical solution that satisfies the functional requirements of the unit."

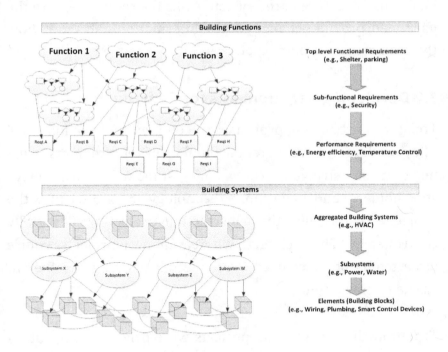

Figure 6: Top-down Functional Decomposition and Bottom-up Assembly of building systems (adapted from [5])

Figure 6 "...assumes a similar functional decomposition. The main functions of a building (such as to "provide shelter" and "facilitate organizational processes") are decomposed (top-down) into lower level functions, such as "provide safety," "habitability," and "sustainability." During this decomposition, one will arrive at functional criteria that can be expressed as explicit performance requirements.

The lower half of Figure 6 shows the aggregation of technical systems from basic elements (e.g., window, door, chiller, finishing) until we reach the level of the standard "building systems," such as lighting, sprinkler, and HVAC. Elements belong to only one subsystem, but there exists a multitude of constraint relationships across elements of a system and between elements of different subsystems. These constraints are related to composition, compatibility, and assembly rules.

7.3 Synthesis of a Systems Solution

Synthesis of a systems solution begins by defining a top-level design solution and then expanding that into the subsystems and components that will satisfy the identified functional requirements. This synthesis process translates the functional design into a detailed design, consisting of an arrangement of systems elements, their decomposition, their internal and external interfaces, and the relevant design constraints.

In construction projects, the architect initially synthesizes possible solutions (concept designs) based on the stakeholders' input and system requirements, while taking into account various constraints, such as space and building codes and standards. This synthesis process will evolve from concept design into preliminary, detailed, and final designs, with considerations for alternative configurations and technology options at each step as ways to meet the project requirements.

7.4 Trade Studies

Making trade-offs is critically important in construction projects. These trade-offs are made using trade studies that are used to determine the optimal course of action when designing a system. During the design process, many decisions must be made concerning different options and alternative approaches in the design. There can be a number of trade studies conducted throughout a project to weigh the merits of different solutions, different technology applications, and different deployment options.

One particularly useful and familiar type of trade study is the cost/benefit analysis. Cost is often a primary factor in determining the best solution, so a comparison of each solution's cost weighed against its benefit may be needed.

Another type of trade study often used by organizations is the make-buy analysis. This trade study is most commonly used when the procuring organization itself has some ability to build parts of the system.

Gap analysis is another type of trade study that evaluates an existing system's capability to satisfy the needs of the systems being designed. All trade studies should be thoroughly documented, showing a summary of the benefits, limitations, advantages, disadvantages, and alternatives, and trade-offs that were considered.

7.5 Validation and Verification

In traditional construction projects, requirements are typically verified through the project commissioning process that occurs at the end of the construction phase. Any issue that may arise through this commissioning process would potentially be more difficult to resolve. It is beneficial to employ an SE approach to validate and verify construction projects to help detect issues earlier in the integration process, making it easier to fix those issues.

Validation is the process of proving that the system accomplishes or is able to accomplish its purpose. Validation typically occurs at the end of development when the systems is actually built and operational, because validation can only be accomplished at the system level.

Verification is the process of proving that the design is properly built and integrated and that it is in compliance with the specifications. Verification must be accomplished throughout the entire development process. Verification of each subsystem's performance and compliance is accomplished as the various subsystems are completed.

7.6 System Operation & Maintenance

Reliability, Availability, Maintainability, and Safety (RAMS) analysis conducted during the development of the product/system affects the activities during the operations and maintenance phases. These analyses get translated into requirements for development, are used to make the best decisions during design, and ultimately determine the level of reliability, availability, maintainability, and safety during operation and maintenance. The analyses conducted and the choices made also help determine the operational requirements, the training needed, and the inspection and maintenance practices after the system is fielded.

7.7 Deactivation and Disposal

Deactivation is the decision to remove an older system/facility from the inventory. Disposal is getting rid of the asset. Disposal can be accomplished by recycling, reutilizing, redistributing, or discarding of the project assets. It can be transferred, donated, sold, destroyed, scrapped, or salvaged.

The design team should begin identifying and addressing deactivation and disposal considerations early in the process. Construction projects must disposed of in accordance with legal and regulatory requirements related to safety, security, and the environment.

8. Zero-Energy Building Case Study

Application of the aforementioned SE processes to construction projects is illustrated via the design of a zero-energy building (ZEB). This building is located on the campus of the Florida Institute of Technology (FIT), in Melbourne, Florida. The building design consists of high energy-efficiency technologies integrated with onsite solar energy and a smart building automation systems to achieve net-zero energy consumption. This project is funded partly by the Florida Department of Agriculture and Consumer Services (FDACS), Office of Energy [11].

8.1 Overview of ZEB Design

As defined by the US Department of Energy, a ZEB is an energy-efficient building where, on a source energy basis, the actual annual delivered energy is less than or equal to the on-site renewable exported energy [12]. This definition is illustrated by Figure 7.

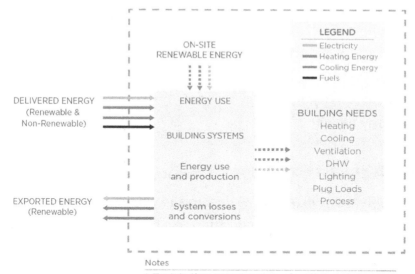

Figure 7: Site boundary of energy transfer for accounting [12]

For ZEB projects, construction cost is undoubtedly the most challenging issue. Selection of a suitable combination of materials, construction methods, and energy improvement technologies is critical to achieve desired energy performance while satisfying cost constraints. Following is the range of typical building design, materials, and technologies to be considered for ZEBs:

- Building envelope technologies, such as roof, wall, and window materials

- Building systems, such as HVAC, lighting, and appliances

- Building designs and automation technologies that integrate occupants' behavior and building operations to minimize operation and maintenance (O&M) costs

- Onsite renewable energy technology to offset any remaining energy needs

8.2 ZEB Project Scope

The scope of this project is to design and construct a commercial office ZEB for which integration of high energy efficiency materials and technologies will help maximize its energy performance. In addition, onsite photovoltaic (PV) solar energy and building load management strategies will be employed to ensure the building will meet it energy consumption target. A longer term objective is to use this project as a guide for building designers and owners to conduct feasibility studies of zero-energy building concepts.

8.3 Project Requirements

Because of time-varying nature of energy resources and building loads, ZEBs are "living systems" that need to continually meet the goal of zero energy on a yearly basis. It is important to define the primary functionalities of the building based on energy efficiency objectives. Requirements engineering plays a significant role in the design of ZEBs.

No.	Primary Requirements	Derived Requirements
1	The building shall be constructed to achieve zero-energy performance.	
2	The project shall establish the design methodology, material procurement strategy, and construction method to achieve zero-energy for new commercial buildings.	1) The project team shall develop and generate a formal report detailing the design methodology, material procurement strategy, and construction method to achieve zero-energy for new commercial buildings located in the state of Florida. 2) The project team shall conduct an energy audit of the completed facility.
3	The project shall specify energy efficiency improvement methods for existing small (< 5,000 ft²) to medium (between 5,000 ft² and 50,000 ft²) commercial buildings located in the State of Florida.	The project team shall develop and generate a formal report detailing energy efficiency improvement methods for existing small (< 5,000 ft²) to medium (between 5,000 ft² and 50,000 ft²) commercial buildings located in the state of Florida.
4	The project shall integrate electric vehicle charging stations into the building electrical power distribution network.	1) Demonstration facility shall provide a minimum of 2 charging stations for electrical vehicles as part of the facility power distribution system. 2) Each EV charging station shall be capable of **TBD** Volt-Amp
5	The project shall provide public access to the demonstration platform and project website to increase awareness and promote understanding of zero-energy building technologies.	1) The project team shall develop a website accessible to the public. 2) The project website shall include information that increase awareness and promote understanding of zero-energy building technologies.
7	The project shall produce a toolbox that can be used by building designers and owners to conduct feasibility studies of zero-energy building design options.	The project team shall develop a toolbox in **TBD** that allows quick feasibility studies of zero-energy building design options.
8	The completed facility shall have automatic control of heating, ventilation, and A/C (HVAC)	1) The new HVAC system shall include a controller capable of monitoring and control inside temperature, airflow, and humidity. 2) The inside environmental conditions (temperatures, airflow, humidity, etc.) as well as occupancy shall be continuously monitored by a network of sensors.

Table 1: Sample of system functional requirements

For the FIT project, the SE team initially established a set of primary functional requirements based on stakeholders' needs, such as building architectural features, occupancy layout, and energy loads. Additional primary requirements were subsequently developed from zero-energy performance criteria. From this initial set of functional requirements, other requirements (secondary) were derived. In systems engineering it is important that all requirements are traceable from a stakeholders' needs. Table 1 illustrates a subset of the system's primary and secondary requirements for the ZEB.

8.4 Concept of Operations (CONOPS)

A concept of operations (CONOPS) is a tool used throughout the SE lifecycle to communicate operational needs, desires, visions, and expectations of the user to designers, integrators, funding decision makers, and other stakeholders. A brief CONOPS of the FIT ZEB follows.

8.4.1 Overview

The expected occupants of the building consist of 5 to 6 full-time employees of the FIT Alumni Affairs, 2-3 part-time members of the energy research team, in addition to transient guests and occasionally scheduled visitors from local communities. The building will operate on a net metering agreement with the local utility company (FPL) to export any excess electricity produced by the onsite PV solar facility.

The building automation systems will be connected to the FIT campus-wide management system. During the testing phase (one year after construction), the building will be operated by the research team with the help of the FIT Facilities department. Once the testing phase ends, FIT Facilities will take over the operation and maintenance for the remaining life of the building.

8.4.2 Normal Operation

The building will operate from 8 AM–5 PM, five days a week, and will host 5-6 users during those hours. During that time, those users will occupy their personal office spaces and need proper lighting and access to computers, Internet, and phone. During the day, student workers and research team members will come and go and will use 4 computers at the reception desk and in the control operation room.

The front lobby area will have 3-4 monitors that continuously display real-time graphics of building operational data. Other electrical loads include a coffee machine, a printer, and a television for a few hours per day. A refrigerator will also be running 24 hours per day. Occasionally, the building will also accommodate educational tours for K-12 students.

After hours, the large conference room will be utilized by students and other organizations meetings that take place about 12 hours a week. Those groups will need access to the Internet, projector, and lighting. The office space would remain closed during that time. These types of meetings will also occur during the weekends for about 4 hours out of an entire month as well as during the work day 4 hours out of the month.

8.4.3 Emergency Operation

During a regular work day, if a power outage were to occur, the PV solar systems is required to shut down per utility regulation. Critical loads such as emergency and exit lighting, security, and alarm systems are fed by the battery storage system. No other critical loads are needed by the Alumni Affairs staff. During this time, the battery bank will also provide power to the research data collection equipment and the building automation system.

8.5 System Architecture Development

Based on previously defined project requirements and CONOPS, the SE team developed a system architecture illustrating the relationship among building subsystems and their interfaces. This is a departure from traditional building construction projects, where there is minimum interconnection among building systems. System architecture serves an important role in ZEB design by providing an insightful framework facilitating the development, configuration, operation, and maintenance of the building systems. Figure 8 is a high-level architecture of the FIT ZEB.

8.6 Trade Studies

It is important to conduct trade studies on ZEB envelope materials and heating, ventilation, and air conditioning (HVAC) systems, since they play a major role in energy savings. These trade studies help optimize the design of the building while also meeting the project's cost requirements.

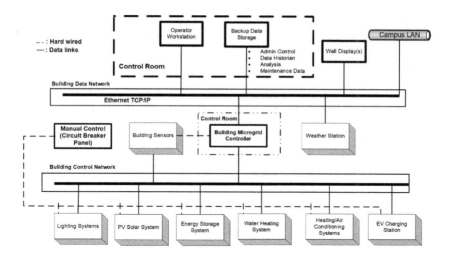

Figure 8: High-level architecture of the FIT ZEB

For this case study, an example trade study for building technologies is shown in Table 2 below. The first step was to identify the energy savings associated with different technologies. Once they were evaluated, the team focused on the combination of technologies that would make the most impact on the building energy performance.

8.7 Application of MBSE in ZEB Case Study

MBSE supports different project organization paradigms. It gives the systems engineer the ability to impose constraints on the direction and type of relationships among the model elements, which are associated with project stages and deliverables. For example, a package may be associated with user requirements. A deliverable could be a package of system requirements with coverage metrics to the user requirements.

Building Category	Technology Type	Deploy Method	Deploy Where	Energy Savings	Advised for FIT Project	Advised for Florida Climate
Building Envelope	Applied Solar Control Retrofit Films	Retrofit	Recommended for single-pane clear glass. For buildings with exposure to direct sunlight without exterior shading, and south and east orientations	Up to 29% HVAC energy savings in warmer climates	Yes	Yes
HVAC	Advanced Rooftop Unit Controls	Retrofit	Commercial buildings currently featuring packaged rooftop heating and cooling equipment with constant speed supply fans.	56%	No	Yes
HVAC	Wireless Pneumatic Thermostats	Retrofit	Any facility with conventional pneumatic controls	Not Reported	No	Yes, for any building with conventional pneumatic controls
HVAC	Smart Ceiling Fans White Paper	New Buildings/ Retrofit	Open office floor plans	4-11%	Yes	Yes
Lighting	LED Troffer and downlight retrofit kits	Retrofit	Laboratories, office buildings, and other commercial spaces in which fluorescent troffers, linear fixtures, downlights, and other legacy lighting technologies are currently in place.	62%	Yes, if the building utilizes fluorescent troffers, linear fixtures, downlights, and other legacy lighting	Yes, if the building utilizes fluorescent troffers, linear fixtures, downlights, and other legacy lighting
Lighting	Occupant Responsive Lighting	Retrofit	Where utility rates are higher than $0.10 kWh and operating hours are over 14 hours	27-63%	No	Yes
Lighting	Integrated Daylighting Systems	New Buildings/ Retrofit	New buildings or retrofits with LPD greater than 1.1 W/ft² and EUI greater than 3.3 kWh/ft²	27%	Yes	Yes

Table 2: Example trade study for building technologies

In the ZEB case study, a logical architecture of the building was created. One aspect of interest is the building envelope. A high-level representation of the envelope is shown in Figure 9. The software used to build the example model is Cameo Systems Modeler. The high-level view is created in a block definition diagram (bbd).

An internal block diagram (ibd) of the building envelope was created to model the interactions among the elements within the building envelope. The building envelope is composed of elements that are associated with a supplier. This characteristics is inherited by all the realized envelope components.

This initial structure can be considered as the base architecture of the envelope. The blocks presented in the model have a cost rollup pattern applied. The cost is added up automatically from the components and totaled in the upper levels for any number of envelope components (e.g., windows). This is performed internally using recursion on a parametric model. At the top level, a non-functional requirement (e.g., cost) is linked via a "satisfy" relationship to a value property (total cost) of the building envelope. The requirement internally contains a mathematical expression (constraint). When an envelope configuration is tested, the parametric model runs automatically, calculating the cost and evaluating whether or not the requirement is passed.

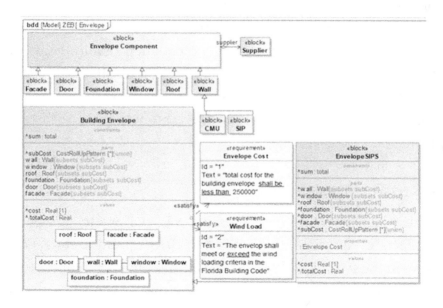

Figure 9: Block definition diagram of the building envelope

The wall component may have two specialized configurations: concrete masonry unit (CMU) or structured insulated panels (SIPS). A specialized configuration is realized by creating an instance of the base configuration. For example, the

121

configuration "EnvelopeSIPS" was derived and specialized for the characteristics applicable to SIPS.

Another example of the usability of MBSE consists of generating the white box interface control documents of the building envelope. By using the pre-determined relationships in the internal block diagram, all the interfacing components can be paired based on its interaction.

All the requirements allocated to the parent block can be mapped to the lower level. If the components are sourced from different suppliers (design and contractors) this will help to model determine if the interfaces meet such requirements. In the case of the ZEB, the SIPS supplier has to discuss how to interface with the foundation, the roof, and the architects to any façade element the need to be attached to walls and withstand wind loads for costal zones.

8.8 System Design

System design is the next step in the SE "Vee" model, driven by artifacts of systems requirements, CONOPS, and architecture. System design comprises several phases: conceptual, preliminary, detail, and final. The ZEB design evolved through these phases, and at each phase, the design was reviewed and updated to reflect changes in requirements or architecture. The final designs are shown in Figures 10 and 11.

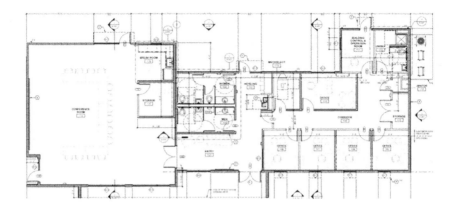

Figure 10: Final architectural floor layout of the ZEB

Figure 11. Final architectural elevation design of the ZEB

8.9 Project Management

The project team members collaborated as three Integrated Product Teams (IPTs), each responsible for a set of highly related project functions and activities. The first IPT is "Building Thermal Envelope & Energy Efficiency," which is responsible for energy audits, thermal envelop design & analysis, energy modeling, and building mechanical systems. The second IPT is "Data Analysis & Facility Demonstration," which is responsible for data mining, analysis tool development, project website development, and project demonstration planning. The third IPT is "Facility Design, Construction, and Integration," which is responsible for site planning, building design, building construction, electrical systems, and systems integration.

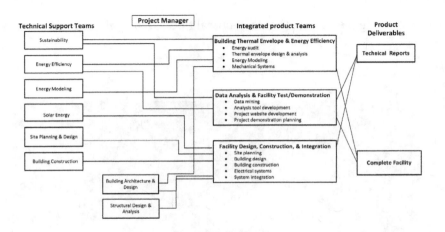

Figure 12: Project integrated product teams

As shown in Figure 12, the three IPTs are supported by a common pool of individual specialists and overseen by a project manager. The collaboration between the IPTs is geared toward successful delivery of the project outputs.

8.10 Procurement Plan

ZEBs are mostly about the annual energy usage, but not the energy required to procure and transport construction materials. A main motivation of this project is to procure materials from local vendors to help reduce the cost of transportation, thereby reducing the overall carbon footprint of the project. As an example, selection of the material for the building exterior walls was made based on its thermal performance as well as its local availability.

The project team also collaborates with local companies for engineering and construction work. The building HVAC ductwork will be designed and fabricated locally by a company in Melbourne, Florida. For other building materials that are not as easy to procure locally, the team investigated alternatives and conducted trade studies to determine whether those alternatives are technically acceptable and available closer to the project site.

8.11 Validation & Verification Plan

A validation & verification (V&V) plan was developed early on in the project, as it is critical to ensure that the building will meet its zero-energy objective. For ZEBs, the inherent interaction between subsystems and human occupancy, along with variation in weather patterns, presents a challenge in the design process, as well as in validating and verifying its intended energy performance. V&V is utilized to measure the effects of changes and to provide data to the stakeholders to ensure the goal of net zero energy is met. The V&V plan includes an energy audit, blow door test, and thermal insulation calculation of the building's thermal envelope through energy modeling and analysis.

The energy audit includes a detailed breakdown of the electrical uses of the building as well as information on the building "shell" (roof, floor, walls, doors, windows). The energy audit identifies and documents total electrical and water usage.

The blow door test (required by Florida Building Code) provides the data on building infiltration (air leakage) through doors, windows, and other openings to support the calculation of HVAC systems energy loss.

Thermal insulation calculations determines thermal performance of the building walls and roof. This calculation helps improve the building energy model. Running these tests helps reduce the risk inherent in the assumptions of a building's performance, which helps make better decisions.

9. Summary

This chapter began by emphasizing the complexity of today's large-scale construction projects and the importance of doing a better, more systematic job of meeting the many stakeholder demands. A short overview of the traditional systems engineering process was provided. The chapter then extolled the benefits of systems engineering and illustrated its much-needed application to construction projects. The steps used in the systems engineering process were tied to the various phases of construction projects in an effort to demonstrate their application to construction projects.

This overview was then followed with a detailed, real life case study illustrating the successful application of the systems engineering approach to the development of a zero-energy building design. Hopefully, the reader has gained a general understanding of the systems engineering process and how it can be applied to construction projects to help ensure the successful development of even the most complex projects.

References

[1]   Bureau of Labor Statistics, "Overview of BLS Statistics by Industry,"   2004.   [Online].   Available: https://www.bls.gov/bls/industry.htm.   [Accessed: 17-Dec-2018].

[2]   ASCE, "2017 infrastructure report card," 2017.

[3]   INCOSE, "Systems Engineering." [Online]. Available: https://www.incose.org/systems-engineering.

[4]   I. Graham, A. O'Callaghan, and A. C. Wills, Object-oriented methods: principles & practice, vol. 6. Addison-Wesley Harlow, UK, 2001.

[5]   M. Jamshidi, Systems of systems engineering: principles and applications. CRC press, 2008.

[6]   H. Lykins, S. Friedenthal, and A. Meilich, "Adapting UML for an object oriented systems engineering method (OOSEM)," in *Proceedings of the 10th International INCOSE Symposium*, 2000.

[7]   H.-P. Hoffmann, "IBM Rational Harmony Deskbook," 2014.

[8]   T. Weilkiens, SYSMOD-The Systems Modeling Toolbox-Pragmatic MBSE with SysML. Lulu. com, 2016.

[9]   A. Aleksandraviciene and A. Morkevicius, Magic Grid Book of Knowledge. Kaunas: Vitae Litera, 2018.

[10]   G. Augenbroe, "The role of simulation in performance based building," Build. Perform. Simul. Des. Oper., pp. 15–36, 2011.

[11]   T. Nguyen, A. Fabregas, and H. Najafi, "Demonstration of a Cost Effective Scalable Zero Energy Commercial Building Design for Florida Climates," 2017.

[12]   K. Peterson, P. Torcellini, R. Grant, C. Taylor, S. Punjabi, and R. Diamond, "A common definition for zero energy buildings," Prep. US Dep. Energy by Natl. Inst. Build. Sci. US Dep. Energy, 2015.

About the Authors

Troy Nguyen is an Associate Professor in Mechanical & Civil Engineering at the Florida Institute of Technology. He spent 26 years leading control & automation R&D projects employing systems engineering methodologies to design and develop mission-critical systems. His current research and teaching focus on construction, energy, and power systems. A licensed general contractor and professional engineer, he has extensive experience in construction projects. He is an INCOSE certified Expert Systems Engineering Professional. He holds a Ph.D. in Mechanical Engineering from Colorado State University (2000). Contact him at tnguyen@fit.edu.

***

Jack Dixon is president of JAMAR International, Inc. He worked for the federal government and the defense industry in project management, business development, and specialty engineering, including ILS, systems safety, reliability, human factors, electrical safety, explosive safety, hazardous materials, OSHA requirements, and electromagnetic & nuclear effects. He is the co-author of *Design for Safety* (Wiley, 2018). Contact him at jamar99@gate.net.

***

Aldo Fabregas is an Assistant Professor in the Department of Computer Engineering and Sciences at Florida Institute of

Technology. His areas of interests involve smart manufacturing, energy systems, and intelligent transportation systems. His research involves systems design assisted with model-based systems engineering (MBSE). He is a certified systems modeling professional. He holds a Ph.D. in Industrial and Systems Engineering from the University of South Florida (2012). Contact him at afabregas@fit.edu.

***

Peter Zappala is the Assistant Director of Student Projects at the Florida Institute of Technology, where he received a dual M.S. degree in Systems Engineering and Engineering Management (2019). He worked on the Zero-Energy Building project at FIT, focusing on energy efficiency and renewable energy technologies Contact him at pzappala@fit.edu.

# # #

A rainbow breaks through the clouds behind the massive Vehicle Assembly Building at NASA's Kennedy Space Center (KSC) in Florida. (NASA/Jim Grossmann)

INCOSE SEP Cohort Guide

Steve Ratts

....

Abstract

Obtaining an INCOSE Systems Engineering Professional (SEP) certification is an excellent way to objectively validate your systems engineering knowledge and experience; but it's also a non-trivial task. One way to make the process easier and more rewarding is to join a cohort of like-minded systems engineers sharing this common goal. This chapter summarizes the experiences and lessons learned by the INCOSE Space Coast chapter organizing and running successful cohorts studying the fourth edition of the systems' engineering handbook. It's intended as a guide to prospective cohort leaders seeking to organize and conduct their own SEP certification efforts.

1. Introduction

INCOSE advises that the average time spent to obtain a Certified Systems Engineering Professional (CSEP) certification is approximately 200 days, or 6.6 months. Obtaining an Associate Systems Engineering Professional (ASEP) certification will require less time primarily because there is no need for references or review of an applicant's experience; even so, ASEP applicants face the same exam and so can expect a similar effort preparing for it.

One way to make the process easier and more rewarding is to join a cohort of like-minded systems engineers sharing this common goal. Working with such a group offers a number of benefits, including exposure to a wider set of perspectives and experience, collaboration on development of cohort training materials, fostering a sense of accountability to help individual members keep each other on track while progressing toward the goal, and of course the comradery that comes from working together on a team.

Some organizations have established cohort processes where experienced leaders can be engaged to organize and lead a cohort. The processes may be formal or informal and are typically put in place through an organization's systems engineering (SE) Community of Practice or other similar network of systems engineers. Alternatively, you may find established cohort processes and experienced leaders available through your local INCOSE chapter, which has a natural interest in promoting INCOSE SEP certification. Wherever you find cohort processes you may also find a network of systems engineers willing to share their knowledge and experience in running such a cohort.

This chapter stems from experiences and lessons learned by the INCOSE Space Coast chapter organizing and running successful cohorts studying the fourth edition of the SE Handbook [1]. It's intended as an aid to prospective cohort leaders seeking to organize and conduct their own SEP cohorts. Topics covered include planning a cohort, promoting a cohort, application preparation and review, and preparing for the INCOSE SEP exam. This guide may be used as a supplement to any other cohort processes you have available or may be tailored to meet the specific needs of your organization.

The following key points are central to this guide, and will aid any cohort seeking to help its members achieve INCOSE SEP certification:

- The purpose of the cohort is to aid its members in preparing their applications as well as preparing for the INCOSE SEP exam. It is NOT the purpose of the cohort to teach anyone systems engineering; thus, cohort members are referred to as members, not students, and each member is expected to contribute to the cohort.

- Cohort leaders act in the role of guides and/or coaches for the members, sharing wisdom, perspective, and personal experience from the point of view of one who has already completed a CSEP or higher certification. They take the lead in presenting on the topics of application and exam preparation; but take a back seat while allowing the cohort members to present material on the INCOSE SE Handbook they've been studying.

- The best way for a cohort member to prepare for the exam is via a "hands-on" approach in which they prepare their own materials. It's not the role of cohort leaders to spoon-feed members training material, and to do so would rob them of the opportunity to best prepare for the exam. Each cohort member must therefore take personal responsibility for producing study materials – the development of which is instrumental in their own preparation for the exam.

- Understanding how INCOSE produces exams, and in particular how real exam questions are constructed, is essential to preparing for the exam. INCOSE follows specific rules when constructing actual exam questions, and understanding those rules directly aids members in visualizing and anticipating the exam.

- Reading the INCOSE SE Handbook is a necessary but not sufficient part of exam preparation. Each member must study the handbook in search of possible or likely exam question material. To this end, a central part of the cohort process includes each member producing their own set of hypothetical test questions, complete with correct and incorrect response choices that adhere to the same rules INCOSE applies to real test questions. This is an essential part of visualizing and anticipating the test. Obtaining sample test questions from any other source circumvents the primary benefit of reading the INCOSE SE Handbook in search of likely question material. The leader's role in this is merely that of a coach assisting in quality control.

- It's not possible to memorize the INCOSE SE Handbook, and trying to do so would not be an effective use of time. It's much better to study the handbook in search of patterns, principles, concepts, and relationships that INCOSE consistently applies. The purpose of the member-led presentations is to highlight likely exam question material and to present and discuss these aspects – not to teach each other SE principles. It's therefore essential that each member has read the material prior to the session so that they can effectively contribute to the discussion.

1.1 Certification Benefits for the Individual

Acquiring an INCOSE SEP certification is not an easy or trivial task, so why do it? To understand the answer to this question, it helps to first define what INCOSE SEP certification is and is not. First of all, INCOSE SEP Certification is not a license to practice systems engineering, nor is it a requirement for employment – although it may be used by some organizations as a qualifier in placement. Essentially, INCOSE SEP certifications offers an objective assessment of an individual's knowledge, education, and experience as these relate to systems engineering, particularly with regard to the ISO/ISE/IEEE 15288 standard. As such it offers a number of potential benefits for individuals, teams, and organizations.

For individuals, obtaining and maintaining an INCOSE SEP certification may provide the following benefits:

- Formally and objectively recognizes your systems engineering knowledge and experience.

- Provides a portable systems engineering designation that is recognized across industry domains.

- May provide a discriminator for job applicants.

- May provide a competitive advantage in your career.

- Helps to indicate your commitment to personal development through the requirement for participation in continuing education.

Essentially, for the individual, INCOSE SEP certification helps to set you apart from your peers lacking certification. There is an additional benefit which may not immediately be apparent. In general, each individual's personal knowledge is concentrated in the areas where they've spent the most time working or studying. This may result in systems engineers with an incomplete baseline of knowledge. The process of preparing for and successfully passing the comprehensive INCOSE SEP knowledge exam ensures each person doing so has a more complete baseline of knowledge. So, while it's not the purpose of the cohort to teach members about SE, it's also not unusual for members to come out of the process with a wider and more complete perspective on SE.

1.2 Certification Benefits for Teams and Organizations

INCOSE SEP certifications also benefit teams whose members hold them. This effect is primarily observed through better communication and coordination within the team and across geographical, organizational, and cultural boundaries. As INCOSE SEP certification is rooted in an understanding and application of the ISO/ISE/IEEE 15288 standard, having team members who are certified helps to ensure common definitions and understanding for SE terms and concepts, which in turn fosters better communication within the team. It likewise helps to break down cultural, organizational, and geographical boundaries within the team.

Finally, INCOSE SEP certification also benefits organizations which employ certified SEPs. Just as INCOSE SEP certification provides formal recognition for individuals, this recognition may be reflected in the assessment of an organization's professional staff. In this way, it can help to provide a discriminator in proposals. INCOSE SEP certification can also impact the hiring and promotion process by providing an independent external assessment. It also helps to motivate employee participation in continuing education – something that ultimately pays dividends in the strength and capability of the organization.

1.3 Why Organize a Cohort?

An INCOSE SEP training cohort offers a number of benefits to members seeking certification. Some of the principle benefits include the following:

- Diversity: Cohort members benefit from the diverse SE perspectives and experience shared by other members.

- Accountability: Cohort members help to hold each other accountable and keep each other on track with the process.

- Division of Labor: Cohort members are able to distribute some of the workload.

- Collaboration: Cohort members are able to help review and check each other's applications.

- Savings: Cohort members save approximately $2K verses other exam preparation courses offered commercially.

- Comradery: Cohort members benefit from the sense of comradery and shared purpose in pursuing the goal.

Actively participating in an INCOSE SEP training cohort will enhance each member's understanding of the material and increase the probability that their goal will be attained.

2. How to Organize a Cohort

The process of organizing a cohort includes selecting the leader and drafting the schedule. This is square one in the process and needs to be complete before the cohort can be successfully promoted or begun. The process of drafting the schedule includes planning for cohort activities, such as preparing and reviewing applications, and planning for test preparation. The final step in drafting a schedule is selecting a start date that aligns the schedule with availability for the necessary resources.

2.1 Selecting the Leader

Step one in organizing an INCOSE SEP training cohort is to select a cohort leader. Ideally, this person is already an INCOSE SEP at the CSEP or Expert Systems Engineering Professional (ESEP) level, so that they may share their personal experience and insight into the application process as well as guiding cohort members through their study of the INCOSE SE Handbook. This responsibility can be shared across a leadership team to ease the burden, or it may be undertaken by a single individual who is willing to make the time commitment necessary. It's definitely possible to lead a successful cohort even if the leader is not yet a certified SEP but rather a cohort member – in fact, this is how the author led his first cohort! However, it's highly beneficial to have a leader or leaders with personal experience in the application and testing process. Such leaders are better able to instill confidence in the cohort that they are on the right track, as well as benefiting the cohort with their personal perspective and insights.

The cohort leader is responsible for scheduling, promoting, organizing, and conducting the cohort. The cohort leader will be responsible for at least some of the presentations at cohort meetings (specifically those relating to preparing and reviewing applications and general exam guidance). However, it's recommended that the majority of cohort presentations be delegated to the cohort members themselves so as to better engage them and ensure deeper learning.

One very important aspect of the cohort leader's role is as a guide and mentor during cohort meetings. The leader needs to be someone who can help engage all cohort members in group discussions so that the full benefit from diverse perspectives and experience may be gained.

The cohort leader should also personally review each member's application prior to submittal, or coordinate the review of applications with other certified SEP volunteers. The recommended steps for application review are described in detail in the Application Preparation and Review section.

2.2 Drafting the Schedule

The first thing that must be considered when drafting a cohort schedule is the amount of work involved and how this work will be organized in a coherent and achievable schedule. There are two central pillars in the work facing the cohort: preparation of applications, and preparation for the exam. The schedule must allow sufficient time for both, which may or may not proceed concurrently. A sample cohort schedule, including assignments / member-led presentations with a breakdown of handbook material covered in each session, is shown in Table 1. This can be tailored, but all material should be covered.

2.2.1 Application Preparation Planning

The application preparation tasks and events which need to be in the cohort schedule include the following:

- Application preparation training event (approximately one-hour presentation followed by Q&A)

- INCOSE member number due date

- Experience matrix due date (CSEP applicants only)

- Draft application due date

- Reference Requests due date (CSEP applicants only)

- Educational Transcript Requests due date (CSEP applicants only)

- Draft application guided review (approximately one hour)

- Final application review package due

- Final application review complete (one-on-one review with CSEP/ESEP mentor outside of cohort meetings)

- Application submission due date

Week	Topic	Discussion	Assignment Due	Presenter
1	Kickoff	Assign Presentations and HW		Leader
2	Application Prep		INCOSE Member #	Leader
3	Section 1-2	Scope and Overview	Experience Matrix	Member
4	Section 3	Generic Lifecycle Stages	Draft Application	Member
5	Section 4.0 – 4.3	Technical Process Subsections	Reference requests	Member
6	Section 4.4 – 4.8	Technical Process Subsections	Transcript requests	Member
7	Section 4.9 – 4.14	Technical Process Subsections		Member
8	Application Review	Guided Cohort Member Reviews		Leader
9	Section 5.0 – 5.4	Technical Management Processes	Final application review package due	Member
10	Section 5.5 – 5.8	Technical Management Processes	Final application review complete	Member
11	Section 8	Tailoring		Member
12	Section 6, 7	Agreement Processes, Org/Enabling Processes	Applications submitted	Member
13	Section 9	Cross-cutting Methods		Member
14	Section 10	Specialty Engineering	Exam Scheduled*	Member
15	Comprehensive Review	Prep, Application, etc.		Leader

(*) This schedule assumes the standard process. It's also possible to follow an alternate process in which a paper version of the exam is done prior to the application. See the INCOSE website for details.

Table 1: Example cohort schedule

Of these, only the application preparation training event and guided draft application review should be expected to occupy cohort meeting time, as the others represent tasks each member is expected to accomplish between cohort meetings. A thorough application training event may be accomplished in approximately one hour, with time allowed for a question and answer session to follow. Similarly, the draft application review can typically be accomplished in one hour regardless of the number of cohort members if the process defined the Guided Application Review section is followed.

The application training event should be led by someone who has personal experience completing a successful application and is familiar with the current INCOSE forms and systems engineering experience areas. This may be a cohort leader, or an outside expert brought in for the purpose. If an expert is not available within your organization, the INCOSE Certification Program central office can suggest speakers on this topic. Email certification@incose.org with a request.

2.2.2 Test Preparation Planning

The majority of cohort meeting time will be devoted to test preparation. Central to this task is completing assigned reading by each cohort member prior to the meeting. Thus, the definition of material to be covered in each meeting will drive the number, length, and cadence of the meetings – and thereby the length of the cohort process. Version 4 of the SE handbook is 241 pages long, and since any part of the handbook may be covered on the exam, all parts should be read and reviewed by the cohort. The example schedule shown in Table 1 breaks down reading assignments and member presentations so that they span approximately 20 pages (17 – 24) in each session, except for the last two which both span 31 pages. This may be done if, by that point in the schedule, the application prep and review tasks are complete, allowing more time for reading and preparation of those member-led presentations.

It should be noted that the comprehensive review may follow at any time after the last member-led presentation, but should occur shortly before the date for scheduled exams. This review is typically at least a day-long event in order to allow sufficient time to cover all sections of the handbook.

2.2.3 Selecting a Cohort Start Date

The old adage "There's no time like the present" applies to starting many tasks; however, it may be less useful for determining when to kick off a cohort. Planning an INCOSE SEP training cohort may begin at any time, but the cohort itself will benefit from more optimal timing of the schedule. With regard to this, there are several things that should be considered beyond mere availability of resources when planning for promotion of the cohort and selecting the kickoff date.

<u>Align Schedule with Organizational Goal Planning</u>

Cohort participants are generally also members of organizations that employ systems engineers, and those organizations may undergo annual goal planning for their members. Such goal planning typically includes establishing time-limited specific goals for each individual that are relevant to their functional area and role in the organization. Such goal planning efforts may be accompanied by encouragement from organizational management for each individual to consider their goals and how they impact their career development. Also, organizations employing goal planning typically also follow up by requiring their members to document their accomplishments in terms of meeting personal goals.

If the promotion and kickoff of the cohort is aligned with an organizational goal planning effort, then cohort members may be encouraged to include INCOSE SEP certification as one of their professional development goals. When cohort members take on INCOSE SEP certification as part of their documented professional development goals at work, this fosters a deeper commitment to the successful accomplishment of the goal. Since goal planning typically takes place on an annual cadence, this also permits sufficient time for the cohort members to complete this goal within a single goal/accomplishment cycle. Thus, cohort members taking on this goal are more committed and thereby more likely to complete it successfully.

Avoid Prolonged and/or Repeated Breaks in the Schedule
Given the length of the cohort process, it's wise to take into consideration major national holidays or other anticipated schedule brakes that may be necessary. For example, in the United States, it's unwise to plan for the cohort to begin right before Thanksgiving, which falls on the fourth Thursday in November. To do so will rob the cohort of momentum when members lose availability due to travel or other commitments and will furthermore set the cohort up for an even larger break in momentum during the Christmas and New Year's holidays.

Moreover, each cohort member's employer may be placing additional demands on their time as those organizations seek to ensure they meet their own year-end goals. Thus, it's unwise to schedule a cohort to start near the end of a calendar year. This dovetails nicely with the above constraint for aligning with goal planning efforts since generally those efforts take place early in the year.

Target a Paper Exam Date and Work Backwards

If the cohort is planning to take a scheduled INCOSE paper exam, then the start date for the cohort may be determined by selecting the paper exam offered by INOCSE and working backwards from that date to determine when to start.

***

Ultimately, any start date that allows for a majority of cohort members to participate without extended breaks disrupting momentum, and which allows the cohort to finish within a goal planning/accomplishment cycle is acceptable and may be considered. Other constraints that should also be considered are suitable venue availability and the availability of the cohort leader(s). If the cohort has a leadership team, they may be able to cover for each other when outside commitments interfere with availability for particular leaders; however, if there is a single cohort leader then that person's schedule and availability must also be accounted for.

3. How to Promote a Cohort

Once a cohort has been planned and the schedule has been established, the leader(s) may wish to organize a promotion effort if they do not already have a waiting list of members wishing to participate. The cohort will benefit from having an adequate size as this results in a greater diversity of perspective and experience when the cohort is presenting and discussing topics. The maximum size for a cohort should allow for each member to be responsible for at least one of the member-led presentations. The minimum size is such that no single member has to present more than three times as preparation for such presentations is a significant burden – typically this means the minimum cohort size is about four.

3.1 Pros and Cons of Limiting Participation

When promoting a cohort, the leader(s) need to consider if the cohort will be limited to only members within a single organization, or if it will be open to any prospective applicant in the local area. This choice impacts how the cohort will be promoted as well as what venues are available and suitable for the cohort to meet at. There are pros and cons to each of these options, some of which are outlined below.

3.1.1 Limiting Participation to Members of a Single Organization

If the cohort is restricted to only include INCOSE members from a single organization, that may have the following benefits:

- Venue provided by the organization (e.g., a conference room with projector and screen at a corporate location)

- Common goal planning/accomplishment cycle for all members

- Access to proprietary organization-provided training resources for all members

- May allow meeting times during normal business hours (e.g., lunch time) if members offset their time (consult management for approval)

However, this option may also limit the pool of available cohort members, and thereby reduce the diversity of perspective and experience spanned by the cohort.

3.1.2 Opening Participation to Local INCOSE Chapter Members

If the cohort is open to any INCOSE member in the local area, this may enable the following benefits:

- Coordinate cohort through the local INCOSE chapter to better integrate members with their chapter

- Greater diversity of perspective and experience spanned by the cohort members if drawn from multiple organizations

- Potentially larger cohort size reducing individual burden of member-led presentations

- Potentially larger and more diverse leadership team

- Potentially larger pool of CSEPs and ESEPs available to for final review of member applications

Of course, this option generally also means that cohort meeting times must be in the evenings or weekends to allow for member's work schedules. Similarly, the venue selection may be complicated as access to corporate-provided conference rooms with presentation equipment is typically limited to members of the organization supplying the venue. Possible venue options in this case may include facilities at public libraries or educational institutions, etc.

3.2 Holding a Promotion Event

Once the decision has been made to limit (or not) the cohort to members from particular organization, and a venue and schedule have been determined, it's possible to schedule a promotion event to get the word out about the cohort and recruit members. This may take the form of a lunch-and-learn or brown-bag session presented internally through your organization, or you may wish to schedule it as a presentation through your local INCOSE chapter – or both.

If you choose to advertise the cohort, you should plan for this event several weeks ahead of the cohort kickoff date to allow interested potential members to consider their own schedules and whether they've got the time and energy available to participate fully.

Any promotion event should cover, at a bare minimum, a presentation detailing the value of INCOSE SEP certification in addition to outlining the schedule and expected work load needed from members. Without a solid understanding of the value of SEP certification, it will be difficult to recruit committed cohort members. INCOSE has a ready-made presentation (incose-sep-overview.pptx) that may be downloaded from the INCOSE public website. The presentation is an excellent basis from which to develop a cohort promotional presentation tailored to the specific details necessary for your cohort.

4. How to Conduct a Cohort

Conducting a cohort involves several steps. A kickoff meeting should be held to set the tone for leaders and participants. Exam applications must be prepared and reviewed. Finally, preparation for sitting the actual INCOSE SEP exam is itself a multi-step process.

4.1 Kickoff Meeting

The cohort kickoff meeting is an important meeting that must include a presentation by the cohort leader(s) informing each member of the cohort schedule and the expectations for their participation. This includes reading assignments, homework expectations, application preparation, and the assignment of member-led presentations. The purpose of this meeting is to get the cohort organized and focused on the tasks ahead.

As the kickoff meeting defines the schedule and expectations, it should also include a presentation by the cohort leader(s) providing an overview of how to prepare for the INCOSE SEP exam. It should be stressed during the kickoff that the purpose of the cohort is to prepare members for the exam, not to teach them SE. Thus, understanding what to expect on the exam and how to prepare for it is essential to each cohort member's ability to produce their required homework questions and prepare for their member-led presentations. This is covered in greater detail in the sections on Homework and Preparing Member-Led Presentations.

4.2 Application Preparation and Review

It's not unusual for applicants to have more trouble with their applications than they do with the actual SEP exam. This is unfortunate, as most problems with applications can be easily avoided. In fact, the simple solution to help prevent many potential application errors is to "do the math."

4.2.1 Experience Matrix

If the applicant is applying for CSEP, then the first step is to construct an experience matrix. Even if only applying for ASEP, this is still a recommended step as the experience matrix will help determine how close an applicant may already be to the CSEP level, as well as where they may wish to invest further effort in their career as they progress towards CSEP.

Fundamentally, the experience matrix maps time spent in various positions and roles (work experience) to the SE experience areas recognized by INCOSE. Furthermore, it also helps to identify and record potential references that will be needed by CSEP applicants when they complete their applications.

One of the common errors made in CSEP applications is inadvertently claiming more time spent in SE areas that can be accounted for in the corresponding position from which the experience is derived. For example, if an applicant spent six months in a given position where they spent roughly 50% of their time working on Requirements Engineering and 50% of their time working on systems Integration, then they would be able to claim only three months of experience for each – not six months each. Simply stated, the total time claimed in SE experience areas may equal but must not exceed the total time in the corresponding position. To claim otherwise would be false and will certainly get an application rejected – requiring the applicant to correct it and submit again, possibly having to restart and repay a new application fee.

Another common mistake applicants may make is head-to-foot counting errors. These occur when the applicant has inadvertently double-counted a period in time. For example, suppose the applicant worked in a position from January through December of a given year. If the applicant lists their next position as starting in December, this effectively double counts the month of December. Double counting periods of time like this will also result in an application being rejected.

To help prevent these and other common arithmetic errors, it's best to construct the experience matrix in a spreadsheet or similar tool where testing for such errors may be automated and potential problems highlighted. As such tools are easy to develop and of great use to all applicants, it's not unusual for SE organizations to produce and maintain them; however, if such a tool is not already available through your organization it's a fairly simple matter to construct one that may be distributed as a template. An example is shown in Table 2.

	P1	P2	P3	P4	P5	P6	P7	
Position								
Title								
Start Date								
Finish Date								Total
Reference(s)								Months
Total Months	0	0	0	0	0	0	0	0
Requirements Engineering	0	0	0	0	0	0	0	0.0
System and Decision Analysis	0	0	0	0	0	0	0	0.0
Architecture/Design Development	0	0	0	0	0	0	0	0.0
Systems Integration	0	0	0	0	0	0	0	0.0
Verification and Validation	0	0	0	0	0	0	0	0.0
System Operation and Maintenance	0	0	0	0	0	0	0	0.0
Technical Planning	0	0	0	0	0	0	0	0.0
Technical Monitoring and Control	0	0	0	0	0	0	0	0.0
Acquisition and Supply	0	0	0	0	0	0	0	0.0
Information and Configuration Management	0	0	0	0	0	0	0	0.0
Risk and Opportunity Management	0	0	0	0	0	0	0	0.0
Lifecycle Process Definition and Management	0	0	0	0	0	0	0	0.0
Specialty Engineering	0	0	0	0	0	0	0	0.0
Organizational Project Enabling Activities	0	0	0	0	0	0	0	0.0
Other	0	0	0	0	0	0	0	0.0
Total per Position	0.0	0.0	0.0	0.0	0.0	0.0	0.0	0.0
% SE per Position	0%	0%	0%	0%	0%	0%	0%	0%

Table 2: Example Experience Matrix template

In the Experience Matrix example shown in Table 2, the applicant only needs to enter information in the cells which are highlighted gold. All other cells contain either descriptive text or formulas which are used to calculate the results needed.

Columns P1 through P7 records experience for a single position or role the applicant has spent time in and includes essential information that will be needed by the applicant when they complete Section 4 of the INCOSE CSEP application (Form 1).

The first block in each column collects information that defines the position. The fields and their definitions and requirements are listed below.

- Position: Applicant's position in this role (e.g., Chief Engineer, Systems Lead, Responsible Engineer).

- Title: Applicant's title or grade within their organization during this role (e.g., Systems Engineer, Senior Systems Engineer, Principle Systems Engineer).

- Start Date: Starting date for this position – note the start date for each position must be greater than the end date for the prior position to avoid head-to-foot errors.

- End Date: Ending date for this position.

- Reference(s): SE references with personal knowledge of the applicant's contributions in this role able to corroborate SE experience claims. It may be necessary to identify multiple references in some positions where no one single reference is able to span the entire period defined for the position.

Between the first and second blocks is a row where a formula may be used to calculate the total number of months spent in each position, given the start and end dates. This is referred to as the Total Months Row and should be calculated automatically by the spreadsheet using the starting and ending dates entered in the first block. This calculated value may be used by formulas in the second block as described below.

The second block in each column enables the applicant to document the fraction of time they spent working in each SE experience area while in that position. This needs to be measured in integer months, and the sum of time spent across all SE experience areas must not exceed the total time spent in the position calculated above in the Total Months Row. It's often difficult for an individual to recall exactly how much time was spent in each SE experience area, however it may be relatively easy to document which areas were worked and to fairly estimate the relative fraction of time spent in each one (e.g., 10% of time spent in X, 25% of time spent in Y, and so forth). If this is the case, then one way to help prevent arithmetic errors is to populate the second block not with integer values but rather with equations which make use of the corresponding value calculated in the Total Months Row. For example, '=0.1*C7' indicates 10% is allocated for this category, where the column for this position is C and the Total Months Row is 7. Adjust as needed for your actual experience matrix spreadsheet.

Below the second block are two rows which may be used to help check the data entered. The first of these is the Total per Position Row, which sums up the total number of months allocated across all SE experience areas for each position. This value may be checked against the corresponding value in the

Total Months Row to ensure the applicant is not inadvertently claiming more experience than is possible in each position. The second check row is the % SE per Position Row. This row may also be used to check the applicant's entries in the second block and should never exceed 100%. If the applicant happens to know that only a percentage of their time was spent working in SE experience areas due to other non-SE responsibilities, then this row is useful in helping to ensure experience is not over or under represented. In such a case, the % SE per Position Row should indicate the correct percentage given the distribution of the applicant's responsibilities.

As the applicant fills out columns P1 through P7 in the experience matrix, the spreadsheet may automatically compile these into the Total Months Column. This is the column shown on the far right of the experience matrix and is different from the Total Months Row. Where the Total Months Row recorded the total number of months spent in each position, the Total Months Column records the total number of months across all positions which were spent in each SE experience area. This information is vital in determining if the applicant is in fact ready to apply for CSEP or ESEP.

4.2.2 Application Preparation

ASEP applicants need to complete INCOSE Form 1A Individual Application for ASEP, and CSEP applicants need to complete INCOSE Form 1 Individual Application for INCOSE CESP. These forms are similar in many ways except that the CSEP version includes sections for Education and Experience. INCOSE Form 2 – Instructions for Completing Form 1, contains detailed guidance applicable to both versions of Form 1. As such, each cohort member needs a copy of Form 2 and a copy of the version of Form 1 relevant to the level to which they are applying. The sections composing each of these forms are shown in Table 3.

ASEP: Form 1A	CSEP: Form 1
Section 1: General Information	Section 1: General Information
Section 2: INCOSE Membership	Section 2: INCOSE Membership and Certification Interest
Section 3: Fee Payment	Section 3: Fee Payment
Section 4: Affidavit by Applicant	Section 4: Education
Section 5: Optional Information	Section 5: Experience
	Section 6: Affidavit by Applicant
	Section 7: Optional Information

Table 3: ASEP and CSEP application sections by form

The General Information section records information about the applicant, including contact information, and so on.

The INCOSE Membership section records the applicant's INCOSE member number, and any existing certification information in the case of CSEP applicants.

The Fee Payment section includes questions about the applicant's organization – specifically if there is a Memorandum of Understanding (MoU) or Memorandum of Agreement (MoA) between INCOSE and the applicant's organization, and also whether or not the applicant's organization is a member of INCOSE's Corporate Advisory Board (CAB). Cohort members should check with their organization to see if either of these are applicable to them.

CSEP applicants will need to complete Section 4 of their application, which identifies each institute of higher learning (college or university) they are reporting with details about when they attended and what degree they obtained. Furthermore, they will need to self-identify regarding the relevance of their education (i.e., "Qualifying degree", "Bachelor's degree with no qualifying degree", or "No bachelor's degree and no qualifying degree"). INCOSE will make the final determination whether an applicant's degree is a qualifying degree.

CSEP Applicants will be able to populate much of Section 5 using the information compiled in their experience matrix; however, there will be some additional information they will need to supply. The additional information takes the form of a position summary detailing their role in the systems of interest for each position, followed by summaries of their specific experience in each SE experience area they are claiming for the position.

For ASEP applicants, the Optional Information section includes a place to report education.

4.2.3 Guided Application Review

One very simple and easy way to help make sure each cohort member's application is free from obvious defects and ready for a final one-on-one review with a CSEP or ESEP mentor is to conduct a guided application review. This is an optional but recommended part of any cohort and will typically occupy one entire cohort session, thereby providing a break from presentations and an opportunity for members to catch up on reading assignments if they've fallen behind.

To efficiently conduct a guided application review, each cohort member needs to arrive with a printed copy of their draft application including, if they are a CSEP applicant, a copy of their experience matrix attached to it. They then pass their own application "one to the right" so that each cohort member is now looking at another member's application. This is done for the simple reason that we all tend to overlook our own errors and consequently have an easier time spotting errors others have made.

The cohort leader may then conduct a guided application review by using a blank application to methodically go through each field with the cohort members following along to ensure there are no missing elements in the application they are reviewing. The leader should guide the cohort members in checking the math in the experience section. It may help if cohort members arrive with a hand calculator to enable them to quickly and easily check this math. Math in the application may be checked both directly (making sure claims add up to plausible values), and indirectly (making sure what is claimed in the application matches what is documented in the experience matrix).

In addition to checking the math, each application needs to be checked to ensure that the experience summaries correspond to the documented SE experience areas. If an applicant is claiming experience in any SE area, that experience should be at least mentioned in the corresponding position summary.

Position summaries should also be checked to ensure they are free from discipline-specific jargon or references to secret programs or projects. If acronyms are present in the experience summary, they should only be those defined in the INCOSE SE Handbook. It is recommended that the terminology INCOSE uses to define the SE experience areas be employed in each applicant's position summaries. For example, INCOSE defines systems Integration as "Preparing, performing and managing systems element implementation; Identifying, agreeing and managing system-level interfaces; Preparing and performing Integration; Managing integration results". If an applicant spent significant time working Systems Integration tasks, then they should map their actual work experience to the relevant INCOSE terminology in their performance summary. The definitions for each SE experience area may be found in INCOSE Form 2 – Instructions for Completing Form 1, which each CSEP applicant should have as a reference.

Although experience is obviously acquired by working on programs or projects, details about such efforts which are not directly relevant to SE experience areas should be omitted. INCOSE needs to be able to simply verify that an applicant has experience in the areas they've claimed, but has no need to understand details such as the project name, schedule, budget, or any project or technical accomplishments unrelated to SE. Therefore, project specific details are best omitted from the application.

4.3 Preparing for the INCOSE SEP Exam

Preparing for the INCOSE SEP exam is accomplished primarily by two cohort activities. First are the member-led presentations reviewing assigned material to highlight patterns, principles, concepts, relationships, and likely or possible question material. Second are the member-generated homework questions (recommend five per member per session). Both activities require the cohort members to study the assigned section of the INCOSE SE Handbook.

The value derived from cohort members producing homework questions and member-led presentation stems from the benefit of reading the handbook in search of patterns, principles, relationships, and likely question material. When the handbook is studied in this manner you're actively anticipating what may be on the test. As it's not possible to memorize the entire handbook, and it's also not the purpose of the cohort to teach members about SE, it's vital that the members spend their efforts studying the handbook in such a way as to maximize their likelihood of passing the exam.

One way this is done by focusing the content of the member-led presentations on recognizing and discussing the patterns, principles, and relationships INCOSE applies throughout the handbook. By having each member take personal responsibility for one or more presentations, the work is divided, and the benefit gained from preparing presentations is fairly distributed. Each member-led presentation session should include active participation by all cohort members, so that the diverse perspectives of the cohort may be shared.

The production of member-led presentations and homework both benefit from understanding how INCOSE constructs exam questions, but actually producing homework questions takes this one step further. Each fully-formed question requires a set of possible incorrect responses in addition to the question statement and set of necessary correct responses. Giving thought to likely or possible incorrect options (i.e., "distractors") helps cohort members anticipate and be alert for such things on the actual exam. When this is combined with the search for likely question material you're actively anticipating what the test may be like. This is a technique called "visualization" that has been proven to help people mentally prepare for tests. Visualization helps increase readiness while simultaneously helping to reduce test anxiety.

4.3.1 INCOSE's Exam Question Philosophy

To produce good homework questions, it's necessary to first understand a few things about the goals, guidelines, and processes INCOSE employs when developing actual SEP exam questions.

Fair and Unbiased

INCOSE statistically assesses test results looking for any evidence a given question may be biased for or against different groups. When they find such questions, they either revise or remove them as needed. Fundamentally, they don't want the test to be biased in favor of or against any group – such as more or less experienced engineers or those with or without experience in specific areas. The test is about knowledge, not experience.

INCOSE exam questions are also written and reviewed to ensure they are not biased in favor of native English speakers. It's

INCOSE's intent that any sufficiently knowledgeable English-speaking systems engineer, regardless of background, experience, or native language will be able to pass the exam. This means that each cohort member's question statements should be written to be clear and easily understood by anyone, regardless if they've learned English as their native language or as a second language.

INCOSE also doesn't want the test to be biased for or against different age groups, cultures, geographic regions, genders, etc. Cohort members should keep this in mind while writing their homework questions, and likewise leaders should keep this in mind when reviewing member homework for quality.

Not Too Hard or Too Easy

When INCOSE finds that a particular question is never or seldom answered correctly by anyone, they revise it or remove it. Likewise, when they find that a particular question is too easy or obvious (everyone or nearly everyone taking the exam gets it right), they will similarly revise it or remove it. It's INCOSE's intent that the exam is composed of fair and reasonable knowledge-based questions that effectively test a person's knowledge and understanding of SE as well as their fluency with ISO/ISE/IEEE 15288 terminology and concepts.

This guidance applies equally to member homework questions. The cohort leads should actively watch out for questions that may be obvious or unusually tricky – providing guidance to cohort members that enables them to revise their homework questions accordingly.

Continually Evolving

In their effort to continually improve and refine the SEP exam, INCOSE routinely includes some questions in each test that do not apply to the final score, but from which they assess valuable statistics enabling the questions to be evaluated as candidates for use as actual scored questions in future tests.

In an electronic exam there will be 120 questions, however only 100 will apply to the final score. Similarly, in a paper exam there will typically be 120 questions, only 100 of which are applied to the score. Those questions which do not apply to the score are all ones INCOSE is trying out to see if they are acceptable for future use. They are evaluating them to make sure they are fair and unbiased, not too hard, nor too easy.

You won't know which questions these are, but you can be sure there are questions in your test that INCOSE is evaluating which will not factor into your score. If, in the actual exam, you find a question that seems unusually difficult or unlike the material you've studied, keep in mind that it might be one of these – just answer it the best you can and move on continuing with the test.

Question Format

INCOSE is very specific about the format of the SEP exam questions [2], so cohort members should take this into consideration when creating their homework questions.

- All questions are multiple-choice with either four or five options from which to select.

- No questions are True/False, as there are never fewer than four possible response choices. Never write a True/False question.

- Answers do not reference other answers – there is never an "all of the above", or "none of the above" option.

- Questions and response options will seldom have acronyms in them. If there are acronyms present, they will be from those defined within the INCOSE SE Handbook.

- There will always be either two or three possible incorrect choices presented (which INCOSE calls "distractors").

- If there are five total possible options to choose from, three of these will be correct responses.

- If the number of correct responses required for the question is more than one, that fact will be indicated at the end of the questions (e.g., "Choose 2"). If you do not see an indication of how many options you should select, then there is only one correct option presented.

- Correct responses are never false statements. This means you will not see a question on the exam like, "Which of the following are not true…". Correct responses are always true statements or true associations with the question subject, and so cohort member questions should always be written to request true information.

- Questions are not written with the intent to confuse or trip you up. There are no "trick questions" on the exam, and therefore there should not be any trick homework questions. This point is particularly important. If INCOSE has included a question on the test, then they intend for it to assess knowledge of a subject they believe is important. INCOSE does not deliberately include questions that test for knowledge that is trivial or unimportant to SE. This perspective should be kept in the forefront of your mind when evaluating a prospective homework question.

4.3.2 Identifying Likely or Probable Exam Material

Identifying likely or probable exam material requires cohort members to think like members of the INCOSE Certification Advisory Group (CAG). The CAG is tasked with selecting material from the SE Handbook to be covered in the exam.

While all exam questions tie back to material covered in the SE Handbook, not all material in the handbook will appear on the exam. It's simply not possible to cover all aspects of the handbook in a two-hour test with 120 questions. Therefore, the CAG must identify the minimal baseline of knowledge to be covered by the exam. However, it's worth noting that this group is not the same as the SE Handbook editors and INCOSE Technical Operations leads tasked with writing the handbook – and thus they have their own perspective on what baseline of knowledge is most important to test for.

Prior to writing actual exam questions, the CAG develops Learning Objectives (LOs), which define the set of knowledge they intend to test for. The LOs are then given to SEP volunteers who write the actual exam questions. LOs are written to be comprehension-specific, not detail-specific, in that a single LO may be used to generate multiple questions which all relate to comprehension of the subject. LOs define the set of knowledge the INCOSE Certification Program feels is important for all certified SEPs to have as a baseline. Therefore, the LOs define a filter for which INCOSE SE Handbook material will be covered in the exam.

How can we identify this set of handbook material? Studying the handbook may yield insight into what the writers of the handbook felt is sufficiently important to document, and this certainly relates to what the exam writers can pick and choose from, but how does one get from the set of things documented in the handbook to the subset which is likely to be on the exam? The unfortunate answer is that without inside knowledge, it's impossible to know for sure. However, there are techniques that may help identify likely and useful LOs.

One approach is to see if you can write a good LO for every diagram or figure in the handbook. This is predicated on the assumption that if INCOSE thought a subject deserved a graphical depiction to help convey it, then it may be one that's covered on the exam. For example, suppose you were presenting the figure as a chart in a presentation. What would be your key points and major takeaways? Anything you would highlight in a presentation of that figure is a good place to look for LO material.

Similarly, subjects that are covered using more words or space in the handbook may also be a good place to look for LOs. This is predicated on the assumption that the INCOSE CAG members likely share a fundamentally similar, though not identical, perspective as that of the SE Handbook editors. As with diagrams, imagine you need to reduce a topic into a single presentation slide with takeaway and key points. This may similarly help to identify likely and useful LOs.

4.3.3 Learning Objectives

Learning objectives not only define the set of knowledge the exam will test for – they also guide the selection of correct responses for questions as well as the set of incorrect options ("distractors"). As each LO can potentially be applied to multiple questions, they should not focus on minute details, but rather on significant SE concepts, principles, and processes.

Time spent developing good LOs will benefit the cohort as it results in generating higher-quality questions which will be more like those on the real exam. A well-written LO will make it easier to identify likely exam material, and also to identify good correct and incorrect options for questions. Good LOs also make it easier for the cohort to review questions and collaborate on their development. It's fine to have multiple homework questions that all share a common learning objective – provided the question statements and correct/incorrect response options are different.

<u>Example Learning Objectives</u>

Each learning objective should be concise and specific, defining a knowledge-based subject that is important for a Systems Engineer to understand. Learning objectives may focus on a single topic, the relationship between various things, a contrasting difference between things, or any other knowledge-based subject important to SE.

Below are some examples of learning objectives that may help in drafting others. They have been grouped into categories that characterize the learning objectives in ways that help to define or constrain how to write or select reasonable question statements, sets of correct choices, and sets of incorrect choices. These categories are by no means a complete set of all such possible categories. Cohort members should feel free to develop their own as needed, which the cohort leaders may want to review against the guidance above for INCOSE's Exam Question Philosophy.

- Quantifiable Relationships:

  o Understand the relationship between total program budget and spending on SE.

  o Understand the relationship between the cost to extract defects and the lifecycle stage at which the defect is discovered.

- Qualitative Relationships:

- o Understand the difference between management and leadership according to Kotter (2001).

- o Understand differences between Systems Thinking and Classical Science-Based Thinking.

- INCOSE Terminology:

  - o Recognize the inputs, activities, and outputs of the Design Definition Process per ISO/IEC/IEEE 15288:2015.

  - o Recognize Technical Processes per ISO/IEC/IEEE 15288:2015.

- INCOSE Process/Flow:

  - o Understand the relationship between the Architecture Definition Process and the Design Definition Process processes per ISO/IEC/IEEE 15288:2015.

  - o Understand the relationship between the Development and Production generic lifecycle stages per ISO/IEC/IEEE 15288:2015.

- INCOSE Concepts:

  - o Recognize Habits of a Systems Thinker (Walters Foundation, 2013).

- o Understand the purposes of the generic lifecycle stages per ISO/IEC/IEEE 15288:2015.

Relating a Learning Objective to a Homework Question

The learning objective should be directly related to the question statement and the corresponding lists of correct and incorrect choices in the homework question. How these things relate depends on the type of learning objective. The examples above are decomposed into further specific guidance on how to select question content for each case.

Quantifiable Relationship Learning Objectives

Quantifiable Relationship Learning Objectives seek to assess an individual's familiarity with and comprehension of quantitatively described concepts and how these apply to SE. This is important in that INCOSE seeks to establish a common baseline understanding of these concepts and their importance to SE.

If there is a quantifiable relationship between two things, such as the "relationship between total program budget and spending on SE," then the question statement should ask for something specific and important to this relationship. For example, "According to Honour (2013), what is the optimal percentage of total program cost to be spent on SE in order to minimize program cost," or "What is the cost factor to extract defects when they are discovered during the Design Stage according to the Defense Acquisition University (DAU)?"

If asking for something specific, the list of correct choices for the homework question may be very short or singular (i.e., the optimal value, in this case 14%). The list of incorrect choices should then include plausible erroneous values or false statements, such as "X and Y are independent," "More spending on SE always results in lower program cost," and so on.

Plausible erroneous values should generally be specified to the same decimal place as the correct value (e.g., 10%, 15%, 16%), or should be different from the correct value in such a way that if rounded off to the same decimal place of the correct value the difference is still apparent. For example, in this case, incorrect values of 13.7% or 14.4% would be poor choices since without access to the actual analysis it would be impossible to say they're not the correct answer; but 15.1% is a good distractor since when rounded off to the nearest whole percentage like the correct answer it's still different from the correct answer.

Qualitative Relationship Learning Objectives

Qualitative Relationship Learning Objectives seek to assess an individual's familiarity with and comprehension of qualitatively described concepts and how these relate to each other through contrasting or different attributes. This is important in that INCOSE seeks to establish a common understanding of concepts important to SE.

If the relationship between two things is not quantifiable but may be described qualitatively (e.g., the difference between management and leadership), then the list of correct choices should all be true associations of qualities given the question statement, and the list of incorrect choices are all false associations of qualities for the question statement.

Incorrect choices may be selected from sets of true associations for some other topic so long as they are false for this topic. For example, if the question was, "Which of the following are qualities typical of a leader?" you may want to include typical qualities of a manager as incorrect choices. You may also want to include qualities which are not associated with either a leader or a manager (e.g., "Focuses on individual productivity and contributions").

INCOSE Terminology Learning Objectives

INCOSE Terminology Learning Objectives seek to assess an individual's familiarity and fluency with INCOSE terminology. This is important in that INCOSE seeks to establish a common language of SE vocabulary, concepts, and definitions to foster better communication between systems engineers.

If the learning objective relates to recognizing or using INCOSE terminology (e.g., "Recognize the inputs, activities, and outputs of the Design Definition Process"), then the list of correct choices should be drawn from the set of INCOSE terminology defined by the question statement, and the list of incorrect choices may be either:

- Similar (plausible) terminology from unrelated sets. For example, if correct choices are inputs to the Design Definition Process, then incorrect choices may be inputs to some other process that are not also inputs to the Design

Definition Process, or possibly outputs from some other process, so long as they are not also inputs to this one. In this example, incorrect choices may also be activities within the process, in which case the question will test for understanding the difference between a process's inputs and activities.

- Similar sounding (confusing) but incorrect terminology (e.g., retirement process vs. disposal process)

It is not necessary to include every item in a set when constructing either the correct choices list or the incorrect choices list, provided you have enough choices without the complete set. Keep in mind the learning objective and how well particular choices work in terms of testing relevant knowledge. Watch out for selecting common terms, such as process inputs, that are used in many places. These can produce unneeded confusion due to their association with many things, unless the question is aimed at identifying things which are common in several or many places.

INCOSE Process/Flow Learning Objectives

INCOSE Process/Flow Learning Objectives seek to assess an individual's familiarity with and comprehension of the relationships between SE lifecycle stages and/or between the processes within them. This is important in that INCOSE seeks to establish a common understanding of generic SE lifecycle stages and process which may be tailored through lifecycle model management, etc.

INCOSE Concept Learning Objectives

INCOSE Concept Learning Objectives seek to assess an individual's familiarity with and comprehension of important SE concepts and methods. This is important in that INCOSE seeks to establish a common understanding of these important concepts so that they may be applied correctly where beneficial.

4.3.4 Questions and Correct Response Sources

All exam questions and their correct responses come from the INCOSE SE Handbook. You will never get a question that wasn't covered in the handbook, and there are no correct responses to questions that do not map to corresponding material in the handbook. This means that when constructing homework questions, the question statements and set of correct choices must similarly be from the handbook.

Exam questions naturally also relate to specific learning objectives and are tailored to help test for those objectives.

4.3.5 Incorrect Responses (Distractors)

Incorrect responses are matched up with question statements based on the learning objective and what sorts of distractors are appropriate to use with the correct choices in order to test for this objective.

Incorrect response choices on the actual INCOSE SEP exam are often true statements or real terminology associated with some other subject, which in the context of the given question are false. Such choices may sound true due to familiarity with them resulting from their actual association with some other subject in the INCOSE SE Handbook. When these are present, INCOSE is seeking to test your comprehension and understanding of the larger scope, which would enable you to recognize such things as not being correctly associated with the question at hand.

For example, if a question asks you to identify Technical Management Processes, and one of the choices is the Lifecycle Model Management Process. The Lifecycle Model Management Process is indeed a real process per ISO/IEC/IEEE 15288, but it belongs to the set of Organizational project-enabling processes, not the set of Technical Management Processes.

5. INCOSE Training Resources

There are many resources available to help cohorts and students prepare for the INCOSE SEP exam. At the time of writing, the following are representative of the sorts of training resources available on the INCOSE website:

- https://www.incose.org/docs/default-source/certification/incose-sep-overview.pptx

- https://www.incose.org/docs/default-source/certification/what-it-truly-means-to-be-a-systems-engineering-professional.pdf

- https://www.incose.org/systems-engineering-certification/certification-forms#Application1

References

[1] Walden, David D., et al. *Systems Engineering Handbook: A Guide for Systems Lifecycle Processes and Activities.* Wiley, 2015.

[2] Wright, Courtney. *INCOSE Certification Program Knowledge Exam* (Update 2015). INCOSE International Symposium. 25. 93-111. 10.1002/j.2334-5837.2015.00051.x.

About the Author

Steve Ratts is a Cyber Systems Engineering Manager with Northrop Grumman. He was previously with Raytheon Missile Systems. Having spent 24 years working at defense contractors as a systems engineer, his experience spans all phases of the weapon system life cycle, from concept through production and test. He is an INCOSE Certified Systems Engineering Professional (CSEP), a Subject Matter Expert in Systems Security Engineering, and the current president of the Space Coast chapter of INCOSE. He is passionate about Model-Based Systems Engineering (MBSE) and an advocate for the application of its principles to the systems engineering lifecycle. As the Northrop

Grumman INCOSE Community of Practice Certification Lead, he is responsible for coaching cohort leads tasked with organizing INCOSE SEP training cohorts across the company. He holds an M.S. in Aerospace Engineering from the University of Arizona (1995). Contact him at steven.ratts@incose.org.

# # #

The Falcon Heavy demonstration mission by SpaceX lifts off on February 6, 2018 from historic launchpad 39A at Kennedy Space Center in Cape Canaveral, Florida. The giant Falcon Heavy produced five million pound of thrust, having more than twice the lift capacity of United Launch Alliance's Delta IV Heavy. Amazingly, the rocket's two side boosters successfully returned to Earth automatically, and the rocket's payload (Elon Musk's original Tesla Roadster) was launched into space. (Steve Ratts)

Systems Software Safety

Stacy Strickland

....

Abstract

This chapter reviews the experience of the author with how systems software safety has been approached in some of the aerospace and defense industries on Florida's Space Coast. It provides a brief outline of the theory and practice of systems software safety as used by companies on projects the author became familiar with while working in the systems engineering discipline. The chapter also provides some suggestions for the improvement in the application of systems software safety and how a better understanding of its application may be implemented in the future.

1. Introduction

As safety critical systems in engineering projects continue to increase, and more engineering projects employ systems with software having additional capabilities, the need for better awareness of software safety in systems has become more evident. The continued use of miniaturization and improved processing capabilities is allowing the spread of safety-critical systems from the original application areas of nuclear and defense to new domains of medical device implants, smart vehicles and traffic analysis, drones, miniature robotic systems, upgraded satellite capabilities, interactive virtual environments, and a host of other activities. These systems are becoming more reliant on efficient, reliable, and safe engineering practices as future technological advances and consumer markets will be expected to produce safety-critical applications and products.

Safety is increasingly being considered as a systems problem (Levenson, 1995). Software is becoming more evident to systems engineering as contributing to a system's safety, or as compromising it by placing systems in dangerous states. Systems engineers are discovering that safety-critical systems require a thorough understanding of the role of software in a system and its interaction with other systems. This chapter attempts to explain the current state of software safety interpretation as experienced by the author at aerospace and defense companies on Florida's Space Coast, and considers some possible paths for improvement in these areas.

Section 2 gives a brief review of the theory and understanding of systems level safety and safety hazard determination. Section 3 describes key areas in software engineering for safety that are usually applied and have been applied at Space Coast aerospace companies. They involve hazard analysis, safety requirements specifications with their analyses, designing for safety in systems, testing for those systems, and resources for certification/standards. Section 4 describes lessons learned for future work in areas that may help in improvement of safety for safety critical systems engineering. Section 5 concludes with a summary of three points that the author believes applies to the Space Coast systems safety engineering field.

2. Theory

Hazard analysis is at the core of developing safe systems. System-level hazards are states that can lead to an accident. An accident is an unplanned event that can result in "death, injury, illness, damage to or loss of property, or environmental harm" (Rushby, 1994). The author has witnessed this on projects at various locations. For example, antenna drives not controlled and driven past their limits to destroy mechanisms, satellites driven off course with incorrect mathematical calculations in their code, weapon systems that continued to fire when supposedly instructed by code to cease firing, code control of remote systems that did not perform properly, and so on.

The results of the system-level analysis are used to make decisions as to which hazards to address. Some hazards are found to be avoidable, so they can be eliminated by engineering design (for example, changing the systems design or the environment in which the systems operates), while other unacceptable hazards cannot be avoided and must be mitigated or handled by the system. In that case, systems safety requirements to handle these are then specified.

Further work determines which software components can contribute to the existence or prevention of the hazard. The common methods are fault tree analysis (FTA), failure modes, effects, and criticality analysis (FMECA), and operability analysis (HAZOP) to help in the determination (Maier, 1995). The systems engineer can combine forward analysis methods (identify possible hazardous consequences of failures) and backward analysis methods (determined whether the potential failure is credible to that system). Some of these approaches were originally discussed by Lutz and Woodhouse (1997).

Typically, safety requirements for the software are derived from the resulting descriptions of how the software behaves. The requirements act as constraints on the systems design. Software may be required to prevent the systems from entering a hazardous state (e.g., timeouts), to detect a dangerous state (e.g., overpressure), or to move the systems from a dangerous to a safe state (e.g., reconfiguration).

The design specification is then analyzed to confirm that it satisfies the safety-related software requirements. With installation and testing, verification continues to assure that the design is correctly implemented to remove or mitigate the hazards. The delivered systems is validated against the safety-related requirements with observation continuing during operations to ensure requirements are adequate.

Hazard analysis is iterative, with additional safety requirements being discovered during design and integration of the system. At some companies, this iterative process is not used, for example due to concerns of budget over-runs or schedule slippage. With the systems safety engineer usually of the mindset that there is always enough time and schedule to ensure safety, this can cause some conflicts. Hazard analysis is also used as a best practice at companies for helping prioritize requirements to ensure testing resources are applied to offer the most testing on the greatest vulnerabilities for the system.

3. Practice

This section provides a brief review of several issues that local aerospace and defense companies consider and practice for ensuring safety in software development.

3.1 Safety Requirements Specifications with Analysis

Aerospace and defense companies normally implement formal specifications for safety-critical functions. The notations for this make review, design, implementation, and development of test cases easier and more accurate. Another motivation is that it allows formal analysis to investigate whether certain safety properties are preserved. For example, in avionics, if a backup channel is in control and is in a safe state, it will stay in a safe state. Automated checks that the requirements are internally consistent and complete are often then available (Lutz,2000).

Executable specifications allow the user to exercise the safety requirements to make sure that they match the intent and reality of formal requirements. For example, there was concern with a spacecraft project whether a low-priority fault-recovery routine could be pre-empted so often by a higher-priority fault recovery routine that it would never complete. With requirements formally specified, it could be demonstrated using an interactive modeling tool that the scenario could occur, and the engineers could remedy the problem before implementation.

3.2 Systems Safety Engineering

System safety engineering focuses on the consequences to be avoided and explicitly considers the systems context. Dependability is concerned primarily with fault tolerance (i.e., providing an acceptable level of service even when faults occur). Sometimes, there is no safe alternative to normal service, so the systems must be dependable to be safe. Real-time systems typically must be fault-tolerant and often involve timing-dependent behavior that can lead to hazards if the systems is compromised. This is a particular concern for Range Safety at Patrick AFB.

3.3 Hazards

In hardware systems, redundancy and diversity are the common ways to reduce hazards. In software, designing for safety may also involve preventing hazards, or detecting and controlling hazards when they occur. Hazard prevention design includes hardware lockouts to protect against software errors, interlocks, watchdog timers, isolation of safety critical modules (a favorite of industries on the Space Coast is to ensure the operating systems is not in a safety-critical path), and sanity checks to ensure the software is acting correctly at various stages.

Hazard detection and control include fail-safe designs, self-tests, exception-handling, and warnings to operators. Fault-tolerance mechanisms for detecting and correcting known faults in distributed message passing systems have been developed

(Gartner, 1999). This is a common approach at Space Coast aerospace and defense companies.

3.4 Design Tradeoffs

Design tradeoffs are often made between safety and other attributes. For example, design for fault-tolerant systems can contribute to safer systems (e.g., predictable timing behavior), but they can create additional interactions between components and levels of the systems (e.g., coordinate recovery from a hazardous state), which is not desirable in a safety-critical system.

In addition, sometimes there are ethical/legal/financial decisions involved that are not ameliorated by strictly technical solutions. This may become more evident as time-to market and liability issues become more important in industries on the Space Coast.

3.5 Vulnerability to Simple Errors

Consider the loss of the Mars Climate Orbiter spacecraft (NASA, 1999). The root cause of the accident was a small error (English measurement versus metric measurement). The defect was straightforward, could be well understood, it should have been easy to prevent in design, and should have been easy to catch in testing. In conventional engineering systems design, tolerance is such that being within the specification tolerance is adequate. In other words, small errors in such systems will have small consequences. In software, this is not always true.

3.6 Limited Use of Known Design Approaches

Known good practice design techniques for safe systems are sometimes ignored, perhaps due to budget, schedule, or human factors. For example, a complex system was disabled by entry of several zeros into a data field that caused the database to overflow, which in turn caused the consoles and remote terminal units to crash. Protection against such a bad data insertion was a known design method that was not used.

In another project, the reported corrective maintenance was not to fix the design, but to retrain the operators to bypass the bad data field and change the value if such a problem was seen to occur again. Perhaps by means of capturing the actual costs of such failures, and showing management how a known safe design can prevent it from ever occurring, such disasters could be prevented on other projects.

3.7 Testing

Testing is critically important when it comes to developing safe systems. Since safety requirements often describe unchanging conditions that need to hold in all circumstances, testing tends to verify the fault-tolerant aspects of software. Testing can demonstrate the software responds to anticipated or envisioned abnormal or unusual situations. Test cases can be done for boundary conditions (e.g., startup/shutdown) or anomalies (e.g., failure detection/recovery) that might result in hazards (Lutz, 2000).

Assumptions concerning the environment and users should be considered (but are often overlooked). Sometimes, the point where a hazardous state is entered (e.g., deep space with radiation or thermal changes) are not properly tested prior to operations. Precise environmental modeling can help with this (but is again often overlooked). Similarly, incorrect assumptions about the user or operator of a systems can lead to an unsafe system.

For example, for a past project, the author (along with other engineers) was asked to design an operator user interface at a customer operating site. No meetings with the customer had been previously planned and we had been provided no real details for user needs. It was only when we visited the site that we learned the monitor was going into a rack too high to be visible for most people. Moreover, it was to use a keyboard for entry, with no allowance for a mouse attachment.

Human factors research attempts to establish some accurate assumptions for systems, However, it is usually during actual testing that mismatches are discovered. An often-used phrase heard at the Range for Patrick AFB was: "Test like you fly and fly like you test." This tends to presume a deep knowledge and experience with the application area is needed to determine the distribution from which the test cases should be drawn. But it can also result in operations that can be constrained by the tests.

It is not feasible to test a safety-critical system enough to quantify its dependability. Yet some developers, the author found, tend to only do that testing they are familiar with and fits within their budget and schedule. Measuring and modeling software reliability during testing and operations (for example, error profiling) has been discussed in some textbooks (Voas and Friedman, 1995). The approach is useful because it involves analyzing test logs along with requirements being satisfied.

Different formal methods have different strengths. Flexibility to choose the best-suited method for distinct aspects or phases of a system (with different potential hazard states) can be beneficial. Experience at most companies indicate that the test cases and formal methods tend to follow the confidence the tester has from his/her experience. This is fine, but it would be very helpful if a more diverse approach and development of testing techniques could be reviewed for each situation. Then, a group decision could be made about what test cases to use, where to apply them, and what the different results may mean for each of them.

The improved integration of informal and formal methods can be a positive sign for software safety since it lets developers choose to specify or analyze critical software components at a level of rigor they can select. Formal methods allow demonstrations prior to coding of crucial elements of the specifications (e.g., entry to a certain hazardous state always lead to a safe state).

An advantage of integrating from the safety perspective is that many formal methods have been used for both hardware and software specifications. Critical software anomalies can involve misunderstandings about the software and system interface

(Lutz,1996). Sometimes, the use of formal methods can help bridge the gap of understanding between software and system developers. The author has witnessed meetings where the executable specifications, especially with a front-end that the user can manipulate, was able to allow inspection of assumptions and help discover sometimes latent requirements that may affect safety.

When it comes to testing, there are several areas that could be improved. Firstly, requirements-based testing focused on the links between safety requirements and test cases. That includes better integration of testing tools with requirements analysis and improved test cases for safety-related scenarios.

Secondly, evaluation from multiple sources would involve the combining of testing, mathematical and logical review, and certification of personnel and processes. How to structure and combine these disparate information flows is still not achieved at some companies.

Thirdly, field data are very important for requirements elicitation for future product development, for maintenance required to assure safety of the evolving product line, and for identification of realistic scenarios (Lutz,1996). The author found it difficult at times to secure field data concerning COTS software used in the field. Whether this was a supplier's concern for protecting proprietary data, having no actual applicable data, or rather that the data would be found to endanger their sale was difficult to determine. In some cases, data that was based on old sources or was limited in applying to the project at hand was all that was available.

3.8 Safe Reuse of COTS Software

COTS (Commercial off the shelf) software can be a concern for systems safety. There is a need to assess the COTS product to determine the fitness for an application. There had been discussions noted by the author at some of the companies that it could be suggested that suppliers provide a certificate that effectively guarantees the behavior of their software component. How many suppliers would go out on a liability limb to do this and what would it cost them to do this?

Additionally, the systems and the environment need to be understood to identify when software is being used outside the "operational envelope" for which it was originally designed. Another problem is not just confirming it does what is should, but how to confirm it does not do other things as well. The problem of unexpected behavior can be very serious for safety-related COTS products since there is a need for predictable limited interactions.

3.9 Virtual Environments

The use of virtual environment (VE) simulations to help design, test, and certify safety-critical systems was not much evident at the companies familiar to the author. However, their use to help design, test, and certify safety-critical systems is becoming more evident as industrial users become aware of their increasing sophistication and reducing of costs for testing, at least after initial purchase. Methodologies to support use of VE and standards for tool qualification of VE will certainly help in its adaptation for testing and integration of systems safety features.

3.10 Certifications and Standards

Certification criteria for safety critical systems are both more complicated and less well-defined than for other software. Some of the issues are what standards are appropriate for large, safety critical systems composed of subsystems from different domains. Often such systems contain COTS components or subsystems previously certified under different national authorities that now must be integrated and certified. The author has worked on projects where COTS have been certified by a non-US company but had to jump through hoops to obtain a US or customer recognized authority for certification before the product could be integrated and tested on the system.

Overall, there is criticism of current safety standards. Problems include lack of guidance in existing standards, poor integration of software issues with systems safety, and the heavy burden of making a safety case for certification. Recommendations have included classifying and evaluating standards according to products, processes, and resources and constructing domain-specific standards for products (Fenton and Neal, 1998).

4. Lessons Learned

This section describes what the author sees as lessons learned for Space Coast companies that may offer useful results in the future for those involved in software safety critical systems.

4.1 Translation of Informal Notations to Formal Models

One company reportedly used analysis and simulation of a machine-checkable formal model of requirements for flight guidance mode logic to find errors, many of them significant (Lutz, 2000). One of the directions for future work at the end of the report was "engineers wanted a greater emphasis on graphical representation "(Miller, 1998). Integrating graphical design analysis tools (fault trees) with formal methods can enhance systems safety analysis.

Tabular representation is another informal notation that can be linked to more formal notations and tools. Continued work to support rigorous reasoning about systems initially described with informal notations to help demonstrate the consistency between informal and formal methods would be useful.

4.2 Methodology

There does not appear to be a consistent methodology for using formal methods at the Space Coast companies. This may be partly because ready customization to a project's immediate need drives the use of lightweight (i.e., easy to use) formal methods, and that results so far are primarily case studies.

In addition, studies of which approaches best provide support for safety analyses of evolving requirements, design revisions impacting safety, and maintenance would be helpful for company adoption. Different formal methods have different strengths, so having the flexibility to choose the best method for certain aspects or phases of a systems without additional modeling would be beneficial.

The integration of informal and formal methods is significant for software systems safety because it lets developers choose to specify or analyze critical software components at a level of rigor they select. Formal methods allow demonstrations prior to coding of crucial elements of the specifications (e.g., verifying that key safety properties always hold, or that entry to a certain hazardous state leads to a safe state).

Critical software anomalies often involve misunderstandings of the software/system interface; the use of formal methods may help to bridge this gap. Executable specifications, especially those with a front-end that the user can manipulate, have been found to allow the exploration of assumptions of the systems safety engineer and help to determine latent or potential requirements that could affect safety (Lutz, 2000).

4.3 Safe Product Lines

Concerning the safety of product families, the wish list at many companies is rather long. Some want any analysis that could be run on generic aspects to also apply to derived instances. Some want to certify a set of safety-critical systems all at once or through one agency for all systems. To do this, we would need a much better understanding of the extent to which systems with similar requirements can reuse requirements analyses (not usually present at the companies). Doubtless, it is the variations among the systems (requirements, environment, and platform) and the interactions between these variations that becomes the hardest to characterize, formalize, and verify in terms of safety effects.

4.4 Safe Reuse of COTS Software

For the safe reuse of COTS software, there are two problems. The first is the need to understand how to retrospectively assess the COTS product to determine its fitness for an application. Maybe suppliers would need to provide a certificate that guarantees the behavior of software components; not many suppliers may be willing to do this.

The second problem is how to confirm that the software does not initiate or do other things it is not supposed to do. Additionally, review by the author on some projects at these companies indicated unexpected actions by the software can be a concern for systems safety-related COTS products. There is a safety requirement for limited interactions or dependencies among components of the software components.

4.5 Runtime Monitoring

Runtime monitoring is well-suited to known, expected hazardous conditions when doing software testing on systems. Detection of known faults can involve tradeoffs between increased safety on one hand and increased complexity, decreased availability, and decreased performance on the other hand. In comparison, detection of unexpected hazardous scenarios is more difficult.

The use of remote agents by Space Coast companies to compare a system's expected state with its sensed state and request action if the difference is unacceptable can be useful in the future. Profiling systems usage can identify evolving conditions that may threaten the system, deviations from safety requirements, and operational usage that is inconsistent with safety assumptions.

4.6 Education

There are few courses currently offered in universities on the software engineering of safety. At the graduate level, the courses seem to be often part of the master's program curriculum for practitioners. The focus of such courses tends to be methodological (e.g., how to perform a FMECA) rather than on a scientific and theoretical approach to software safety.

There is a need for courses in safety that build on prior education in fault tolerance, security, systems engineering, experimental techniques, and specific applications domains. At the undergraduate level, student exposure to safety-critical systems seems to be minimal. Unfortunately, the notion that one's own software might jeopardized a system, much less a life, is a novel concept to many students.

There is a need for case-based learning modules to encourage a systems approach to software safety A textbook on software engineering for safety is needed (the author could only find Storey's (1996) but there are probably others available or being developed at the time of this writing). A wider use of popular and more up-to-date accounts of accidents and their causes (e.g., the Patriot Missile systems failure of Gulf War 1991, the Marine Corps MV-22 Osprey crash in 2000) in beginning software engineering courses could heighten awareness that software contributes to hazards.

4.7 Collaboration with Related Fields

Collaboration with related fields could improve overall systems safety. For example, ties between safety and security have begun to be explored more thoroughly as offering productive ways to reason about and design safe systems. These include anomaly-based intrusion detection; noninterference and containment strategies; security kernels; and coordinated response to attacks (faults).

The relationships between architecture attributes and safety are promising, but they are still in their infancy. For example, the safety consequences of flexible and adaptable architectures have not been fully explored. Nor has the evaluation of architectures for safety-critical product families, partitioning to control hazards enabled by shared resources, and architectural solutions to the need for techniques that augment the robustness of the less robust components. For instance, when a safety-critical systems is built using legacy subsystems or databases, operating systems with known failure modes, and COTS components from multiple sources, architectural analysis offers an avenue for safety analysis of the integrated system.

Human factors engineering is an area in which additional research and assimilation of existing results are needed. Better understanding of usage patterns (based on field studies) and formal specification of operator's mental models can yield more accurate safety requirements and safer maintenance. One technique (used at a company the author is familiar with) that merits extension to other domains is the list of design features prone to causing operator-mode awareness errors. The items in such a list can be included in checklists for design and code inspections, investigated informal models, or used in test case generation.

Some other areas of possible collaboration include advances in operating systems (e.g., support for real-time safety-critical applications), programming languages (e.g., safe subsets of languages, and techniques relating programming languages to specification languages), and temporal logics (e.g., reasoning about critical timing constraints).

5. Summary

This chapter described the state of safety-critical analysis in some areas of software engineering as experienced by the author at various companies on Florida's Space Coast. It made some suggestions on how companies may make adjustments to improve in those areas. In essence, they include continued exploitation of advances in related fields to build safer systems, better integration of safety techniques with industrial development environments, and improvements in the teaching of software safety analysis at university level.

For the interested reader, there are several good resources available concerning techniques in software safety engineering. Levenson's book is the standard reference for the field (Levenson, 1995). Another author deals with software safety and reliability (Hermann, 1999). There are extensive resources on the web. They include news-groups, mailing lists, courses, publications, conferences, and groups that discuss software safety in academia, government, and industry.

References

Fenton, N.E. and Neil, M. A. "Strategy for improving safety related software engineering standards." *IEEE Transactions on Software Engineering.* 24(11):1002-1013, 1998.

Gartner, F.C. "Fundamentals of fault-tolerant distributed computing." *ACM Computing Surveys*, 31(1):1-26, 1999.

Hermann, D.S. *Software Safety and Reliability.* IEEE Computer Society Press, 1999.

Levenson, N. *Safeware.* Adison-Wesley, 1995.

Lutz, R.R. "Software Engineering for Safety: A Roadmap." JPL, CalTech. 2000.

Lutz, R.R. "Targeting safety-related errors during software requirements analysis." *Journal of Systems and Software*, 34:223-230, 1996.

Lutz, R.R. and Woodhouse, R. "Requirements analysis using forward and backward search." *Annals of Software Engineering*, 3:459-475, 1997.

Maier, T. "FMEA and FTA to support safe design of embedded software in safety-critical systems." *Proceedings CSR 12th Annual Workshop on Safety and Reliability of Software Based Systems*, 1995.

NASA. "Mars Climate Orbiter Mishap Investigation Board." Phase I Report, November 1999.

Rushby, J. "Critical systems properties: Survey and taxonomy." *Reliability Engineering and Systems Safety* 43(2(:189-214, 1994.

Storey, N. *Safety-Critical Computer Systems*. Adison Wesley Longman, 1996.

Voas, J. and Friedman, M. *Software Assessment: Reliability, Safety, Testability*. Wiley, 1995.

About the Author

Stacy Strickland has worked in systems engineering for over 25 years. He currently works for L3Harris Corporation in Palm Bay, Florida as a systems engineer in the Specialty Engineering Department. He is also an instructor in computer security at Eastern Florida State College. He worked for Boeing Company in HW/SW systems integration on the International Space Station, at Lockheed Martin on modernizing HW/SW systems for the Eastern/Western Range Patrick AFB, and as systems safety engineer and reliability engineer for BAe Systems. He is an INCOSE CSEP, certified SW Architecture Professional from SEI, and a Certified Associate Safety Professional. He was recalled to active duty for Operation Iraqi Freedom and served in Qatar. He retired from the US Navy as a Commander. He holds an MS in computer information systems and an MBA from Embry- Riddle Aeronautical University (1990). Contact him at Stacy.Strickland@L3Harris.com.

# # #

The cruise ship Disney Fantasy in Port Canaveral. The port is the world's second-busiest cruise port, based on passenger volume, behind the Port of Miami. Today's cruise ships are excellent examples of large, complex systems. In many ways, their design and user experience represents the future of systems engineering.

(Peter Titmuss; Shutterstock 1306079959)

About the Editor

Scott Tilley is an emeritus professor at the Florida Institute of Technology, president and founder of the Center for Technology & Society, president and co-founder of Big Data Florida, Senior Fellow at the American Security Council Foundation, past president of INCOSE Space Coast, and a Space Coast Writers' Guild Fellow. His recent books include *Systems Analysis & Design* (Cengage, 2020), *SPACE* (Anthology Alliance, 2019), and *Technical Justice* (CTS Press, 2019). He wrote the "Technology Today" column for FLORIDA TODAY from 2010 to 2018. He holds a Ph.D. in computer science from the University of Victoria (1995). Please visit his author website at http://www.amazon.com/author/stilley.

January 1, 2020: The start of a new year in Melbourne Beach.
(Scott Tilley)

CTS Press

CTS
Press

CTS Press publishes books that focus on the interplay between technology and society. We also work with academics and researchers to publish monographs and essay collections. If you are interested in developing something along these lines, contact us!

CTS Press is an imprint of Precious Publishing. Precious Publishing specializes in taking your writing ideas from conception to fruition. We know that your stories are precious to you, and we'll do everything we can to help see your work published.

All of our books are available online from Amazon.com, usually in both print and Kindle formats. You are the author, we are the editor and publisher, and the world's biggest bookstore is the global distributor.

http://www.PreciousPublishing.biz/ctspress

www.ingramcontent.com/pod-product-compliance
Lightning Source LLC
LaVergne TN
LVHW022310060326
832902LV00020B/3384